What Are We For?

Also by Eleanor Roosevelt

When You Grow Up to Vote
It's Up to the Women
A Trip to Washington with Bobby and Betty
This Is My Story
My Days
This Troubled World
Christmas
The Moral Basis of Democracy
If You Ask Me
This I Remember
India and the Awakening East
It Seems to Me
On My Own
You Learn by Living
The Autobiography of Eleanor Roosevelt
Eleanor Roosevelt's Book of Common Sense Etiquette
Eleanor Roosevelt's Christmas Book
Tomorrow Is Now

What Are We For?

THE WORDS AND IDEALS OF
ELEANOR ROOSEVELT

Eleanor Roosevelt

*Edited and with an Introduction
by Mary Jo Binker*

Foreword by Nancy Pelosi

HARPER ⬤ PERENNIAL

NEW YORK • LONDON • TORONTO • SYDNEY • NEW DELHI • AUCKLAND

HARPER ● PERENNIAL

FIRST EDITION

Designed by Jen Overstreet
Illustrations courtesy of Shutterstock:

Page 7: Fandorina Liza	Page 161: hchjjl
Pages 35, 51, 71, 107, 137, 207: Uncle Leo	Page 175: Evgeniia Andronova
Page 89: Aleutie	Page 193: MicroOne
Page 129: DODOMO	

Library of Congress Cataloging-in-Publication Data has been applied for.

ISBN 978-0-06-288947-8 (pbk.)

19 20 21 22 23 LSC 10 9 8 7 6 5 4 3 2 1

Contents

———

Foreword

ELEANOR ROOSEVELT DIDN'T LOOK BACK, SHE LOOKED FORWARD. She was firmly critical of policy that didn't work for people, but she didn't complain about what she was against—she repeatedly asserted what she was for. That's why we still look to her as a leader, a voice of reason and inspiration. It would be so nice to have her here now! Thankfully, we have her words. They guide our emerging leaders as convincingly as they have influenced previous generations around the world.

Eleanor Roosevelt was the longest serving First Lady in the history of the United States. After she left the White House, President Harry Truman called her the "First Lady of the World." She brought remarkable gifts to public life, most importantly her conviction that politics begins with beliefs. Doing what is right makes for a better public servant—and a better society.

Her own service was based on an ability to reach

Americans. Her husband, President Franklin D. Roosevelt, took office in 1933, and early in his first presidential term, she was asked to write a newspaper column. She agreed. Thereafter, six days a week for almost thirty years, until her death in 1962, she wrote My Day.

She usually started with a quick story about someone she had talked to or something on her mind, but often she ended up giving profound advice. For instance: In December 1941, ten days after the Pearl Harbor bombing, she started by recommending a book she had just started about immigrants in America, saying that she hadn't finished it, "so I can't really discuss it." But then closed her column with some clear advice: "If we cannot meet the challenges of fairness to our citizens of every nationality, of really believing in the Bill of Rights and making it a reality for all loyal American citizens, regardless of race, creed or color, then we shall have removed from the world the one real hope for the future on which all humanity must now rely."

This was politics at its finest. Eleanor Roosevelt believed in true civic action. She insisted that all of us should stand up and be counted, even if standing up and speaking up requires more courage than we think we have. She believed in all people—women and men—and she believed that together we would make a better world.

She knew that we could all learn. She took what she learned in Washington to help write a Declaration of Human Rights for the entire world.

In these troubled times, we all could use Eleanor

Roosevelt's wisdom, every day. I grew up in a family of Roosevelt Democrats. My father was a member of Congress who served most of the time that the Roosevelts were in the White House. So I have always been taught that Eleanor Roosevelt gave voice to the aspirations of all Americans. Read this book and be inspired.

—Speaker Nancy Pelosi

What Are We For?

ELEANOR ROOSEVELT AND THE POWER OF THE INDIVIDUAL

ELEANOR ROOSEVELT'S LIFE (1884–1962) IS LIKE A DIAMOND: multifaceted and brilliant. Over the course of seventy-eight years she managed to marry, raise a family, and pursue careers in politics, journalism, education, and diplomacy. Along the way she transformed the role of First Lady, played a vital part in the creation and passage of the Universal Declaration of Human Rights (UDHR), and became an icon of female empowerment.

Much of what Eleanor did and said was based on two profound convictions: first, that every individual mattered, and second, that every individual had a contribution to make. In her case, she dedicated her life to building a country—and a world—where diversity and inclusion were part of the fabric of everyday life. She began that effort by using her natural curiosity to learn all she could about how her fellow Americans were faring and by sharing that knowledge as widely as possible. As a debutante, she worked with immigrant children in a settlement house. Later, as a teacher in a New York City private school, she routinely took her students to visit places where they could see the impact of poverty and neglect on those who suffered from it. Once she became First Lady, she traveled the country to learn firsthand about and address people's needs during the Great Depression, with particular emphasis on the problems of minorities, workers, women, and youth. During World War II she visited military installations at home and abroad to evaluate the living situation and treatment of military personnel and to advocate for better conditions. Equally concerned with civilian home front issues, she traveled to defense factories and plants to monitor the working environment to learn how work-life issues impacted an overwhelmingly female workforce so she could propose practical solutions to any problems. Despite the fear and paranoia surrounding them, she also learned about and championed the cause of those whom the war had marginalized, especially Jewish refugees and Japanese Americans interned by government order after Pearl Harbor.

After she left the White House, her scope broadened to include the needs and desires of people around the world. As the first chair of the United Nations Commission on Human Rights, she persuaded its members to set aside their political and cultural differences as well as their "personal rivalries" to come together to create the Universal Declaration of Human Rights (UDHR), the cornerstone of the modern human rights movement.[1] Thereafter she wrote and traveled constantly at home and abroad to publicize its principles and the work of the United Nations. Even as her international reputation grew, Eleanor continued to prod her fellow citizens, urging them to make democracy work for all Americans, especially those who were still on the margins of society. Failure to do so, she believed, would have serious repercussions at a time when the United States and the Soviet Union were vying for the support of the emerging nations of Africa, Asia, and the Middle East: "If we cannot prove here that we believe in freedom, in the ability of people to govern themselves, if we cannot have confidence in each other, if we cannot feel that fundamentally we are all trying to retain the best in our democracy and improve it, if we cannot find some basic unity even in peace, then we are never going to be able to lead the world."[2]

What Are We For? The Words and Ideals of Eleanor Roosevelt distills the most important ideas from her life and voluminous written, oral, and audiovisual record—twenty-seven books, more than eight thousand newspaper columns, two hundred–plus magazine columns, almost six hundred magazine articles, nine radio programs, three television programs, and an extensive, varied multiyear correspondence—into a convenient guide to some of her most relevant works. Each chapter is organized around a theme, cause, or idea that she championed.

Besides the sheer breadth of her interests, one of the most striking things about Eleanor's life and thought is her conviction that many of life's problems, whether personal or political, had their roots in fear. Fear robbed individuals and societies of their ability to speak out and act. It was the reason nations stockpiled armaments and closed their borders. Above all fear destroyed the possibility of change and growth. "People who 'view with alarm,'" she wrote pointedly, "never build anything."[3] Instead of giving in to fear, she urged her contemporaries to face it, analyze it, and then act to eliminate it from their lives and the lives of others. "We do not have to become heroes overnight," she wrote. "Just a step at a time, meeting each thing that comes up . . . discovering we have the strength to stare it down."[4]

Eleanor spoke from experience. She started life as a shy, self-conscious child afraid of the dark, of displeasing people, of being unloved. Step-by-step she emancipated herself from her fears and insecurities so that by the time she reached the end of her life, nothing frightened her. Her prescription for dismantling fear (confront it, deconstruct it, and then diligently eradicate it) provides a road map for navigating our own anxious age.

Her unflinching willingness to grapple with pressing questions that still bedevil us also strikes a modern chord. She was not afraid to take a stand on controversial political and social issues such as civil rights, war and peace, birth control, and the role of government in a democracy. Her solutions were invariably grounded in the conviction that individuals in a democracy had to take responsibility for their actions and the actions of their government. Apathy and denial were not options, and staying aloof was "a cowardly evasion." To those who railed against the isms of her time, she asked, "What are we for?"—a question that is as timely as today's tweets.[5]

Lastly, Eleanor's life is a master class in how to combine vision with action. The woman who emerges from these pages is a practical idealist: a woman who knew how to compromise without abandoning her principles, how to disagree without being disagreeable—a skill she honed as a political activist and First Lady and further refined while dueling with the Soviet delegates at the United Nations. She knew, too, how to push for a better future while acknowledging the realities of her time. When she failed, she did not give up. Failure was simply an opportunity to try again. Her persistence was a powerful antidote to the cynicism and apathy of her time and an inspiration to our own.

Ultimately Eleanor believed that individuals and individual effort mattered. She believed, too, that she had a responsibility to work on public issues. She met her obligations in these areas by being curious about people, places, and events, because to her curiosity was the gateway to learning, understanding, and ultimately solving the difficult problems of her time. "It is man's ceaseless urge to know more and to do more which makes the world move," she once wrote.[6] Eleanor Roosevelt possessed that ceaseless urge in abundance. Because she did, she moved her country and the world forward. To read her words today is to marvel at her prescience, to be inspired by her vision, and to be challenged to follow her example.

—Mary Jo Binker

One

POLITICS AND GOVERNMENT

ELEANOR ROOSEVELT FIRMLY BELIEVED that a democratic government offered the greatest good for the greatest number of its citizens. At the same time, she recognized that the effectiveness of any democracy rested on the willingness and ability of individual citizens to support it. Without the participation and oversight of its constituents, a democratic government was doomed to fail. "In the final analysis, a democratic government represents the sum total of the courage and the integrity of its individuals. It cannot be better than they are," she wrote shortly before her death.[1]

Citizenship in a democratic government involved more than just voting on Election Day. It required education and effort. Citizenship meant learning about one's community and ensuring that the economic, social, and political needs of all its members were met. Citizenship also meant educating oneself on current issues, taking a position one way or another, and then working to build public support for that view at the appropriate governmental level. Communicating with public officials and holding them to account for their actions were part of the process as well.

While the importance of individual responsibility in the political process was a consistent theme in her work and advocacy, Eleanor recognized that an individual alone could do only so much. True progress, she believed, could be achieved only if concerned citizens came together to work for change. "A group has solidarity," she once observed. "An individual alone has only his courage and his integrity."[2]

Active, sustained citizenship, she believed, would produce the kind of leaders democratic government required: people who had lived and worked among their constituents and understood their concerns. "The basic success of any politician lies in his ability to make his own interests those of his constituents, so that he merges into the community which he serves," she wrote.[3]

Given her penchant for planning in other areas of public life, Eleanor ironically did not think such leaders could or should plan to pursue a political career per se. She thought it was "much better" if they had jobs or a career they could return to once their time in office ended. Having another occupation would give them the freedom "to do what is right," rather than what was expedient or necessary to remain in office.[4]

A government based on such citizen participation and oversight was bound to be slow, cumbersome, and sometimes messy, but compared with the alternatives, Eleanor thought it offered Americans the best chance for "the peaceful and prosperous conduct of life."[5]

POLITICS

Politics is the participation of the citizen in his government. The kind of government he has depends entirely on the quality of that participation.

(You Learn by Living)

The basis of all useful political activity is an interest in human beings and social conditions, and a knowledge of human nature.

(It's Up to the Women)

I have never had the feeling that anything—whether a judicial system, an economic or a political system—devised by human beings must be accepted as though it had come straight from the Almighty and can be neither changed nor improved.

(My Day, May 17, 1950)

I suggest that young people—particularly young women—come to their political activities from the bottom rather than from the top. If they come from the bottom, there will undoubtedly be times when they will feel that political life is sordid, that human beings are disappointing, that their aspirations and desires are frequently rather low, and it is just as well to realize all this, for no useful work is accomplished until facts are faced and accepted.

(*It's Up to the Women*)

Political bosses and political machinery can be good, but the minute they cease to express the will of the people, their days are numbered.

(*My Day, August 6, 1946*)

It is a pleasant thing for a group to find itself lavishly financed. It is unpleasant to be warned that you must not deviate or the financing will be cut off. But worse, in these days of growing conformity, is the threat that if you deviate in your thinking you will be different from your colleagues, your fellow workers, your fellow students, your contemporaries. And that is intolerable to a number of people. A group has solidarity. An individual alone has only his courage and his integrity.

(*Tomorrow Is Now*)

I have long felt that the same amount [of money] should be spent by both parties, that radio and TV time should be given them free, and an equal amount of newspaper advertising and railroad travel should be allotted in different categories and paid for by the public.

(*My Day, February 13, 1957*)

I hate to see us put so much trust in polls. After all, they don't represent reasoned thought.

(White House press conference, January 17, 1939)

POLITICAL PARTIES

The value of our two-party system lies in keeping an even balance between the two, and not letting either one become too secure at any time.

("How to Choose a Candidate," *Liberty*, November 5, 1932)

It is important always to have a strong opposition party no matter which party is in power, and that is why the issues should be clear.

("If I Were a Republican Today," *Cosmopolitan*, June 1950)

I doubt if there is much use in being a real reactionary. But I think there is a very valuable place to be filled by the honest conservative. He is the balance wheel for those who want to move too fast.

(My Day, March 30, 1950)

I used to have to remind the gentlemen of the Party rather frequently that we Democrats did not win unless we had the liberals, labor, and women, largely with us.

(Letter to Harry S. Truman, June 30, 1946)

I do not blame the Democrats for wanting to force the Republicans into a position where they will do foolish and harmful things, and where they will appear to be what they are—shortsighted and more interested in the well-being of business interests than in the safety of this country in the world.

(My Day, August 20, 1957)

There is plenty of room for differences of opinion on how we shall attain our ends. But there is very little room for differences of opinion as to what our ends shall be.

(My Day, March 6, 1950)

POLITICAL LEADERSHIP

I don't believe politics is a career you can prepare for. In our country, I think, it is much better to have an occupation you enjoy and are successful in. Then, if you have the urge to run for elective office or accept an appointive office, you can do so and feel free . . . to do what is right; and this makes for a better public servant.

(If You Ask Me, *McCall's*, April 1961)

The fear of being thrown out of a job unless he is sure of being able to find something to do, or has already made his place where he can easily return, is bound to affect the usefulness of the public servant. He or she is bound to act with an eye toward reelection or reappointment.

(*It's Up to the Women*)

A politician interested only in his own personal advancement is not only useless as a public servant but he will eventually fail.

(*You Learn by Living*)

Leadership is a stern, demanding role and no person or state can lead without earning that right.

(*The Autobiography of Eleanor Roosevelt*)

A good public servant becomes so at a high cost of personal sacrifice. We need such men; when we find them we owe them our gratitude and, above all, our respect.

(*You Learn by Living*)

Moral leadership is the quality in men and nations which makes other nations and leaders believe that they are not completely self-interested; that they have at heart the interests of other nations as well as of their own.

(*If You Ask Me*, 1946)

It would seem that there is enough dissension among all our people today without responsible lawmakers creating more rage and resentment.

(*My Day*, February 20, 1954)

An important ingredient for the politician is the ability to attract and draw people to him. All political action is filtered through other human beings.

(*You Learn by Living*)

A man or woman in public life must learn to listen to everybody's opinions. They must never be prejudiced or dogmatic, they must keep an open mind, but when they have listened and know what they think themselves, they must have the courage to stand by that. They will frequently be accused while they are listening, of vacillation and weakness, but this will do them no harm and it is far better than being obstinate or trying to act without the full knowledge of the situation.

(*It's Up to the Women*)

The basic success of any politician lies in his ability to make his own interests those of his constituents, so that he merges into the community which he serves. Then and only then can he accomplish what he wishes to do.

(*You Learn by Living*)

Many wise politicians in the past have said that New York and Washington never know what the rest of the country is thinking and so the politicians are right to go home, get in touch with their home people and get the feel of the country. What troubles me, however, is that members of Congress and other political leaders seem not to realize that they go home for a twofold purpose. First, to find out what people are thinking, and next, to enlighten them if they have not heard or understood all the facts.

No politician can be too far ahead of his followers, but he is responsible for the progress of his followers and for their full understanding of questions which he has had an opportunity to study in a way which is not open to every citizen. A Congressman is a moulder of public opinion as well as a representative of opinion, and when he goes home is the time when he can mould public opinion.

(My Day, August 7, 1947)

A man who chooses to hold public office must learn to accept the slander as part of the job and to trust that the majority of the people will judge him by his accomplishments in the public service. A man's family also has to learn to accept it.

(*The Autobiography of Eleanor Roosevelt*)

The President of the United States . . . is, or should be, the great educator of the people, bringing issues to them and explaining the situation.

(*You Learn by Living*)

Someday, perhaps, we will learn that what is really important in a Chief Executive is what he believes in for the people and what his record in public office, or in his field of work, has been. When that happens, our campaigns probably will be much duller, but also much less bitter!

(My Day, November 2, 1940)

Any President frequently suffers from his friends as much as from his enemies, and it is the sense of loyalty and gratitude which often gets men in public life into the greatest of trouble.

(Memorandum for Harry S. Truman, March 1, 1946)

When any public man says that we should consider only our own needs—that we should have done this in the past as well as in the present—it shows how little he understands the magnitude of the world situation. He shows, above everything else, that he has learned nothing from world events.

(*Over Our Coffee Cups*, February 15, 1942)

You cannot have a statesman's vision of the future if you are afraid of the present. We are going to live in a world where people of many races are going to be close to us and are going to have equal economic opportunity whether a small group, temporarily powerful here, wishes them to have it in this country or not.

(My Day, July 5, 1945)

No one, no matter how devoted they were to a man or a leader, can carry on any program by simply doing what they think that man would have done. It is impossible to tell how any man would have changed his plans or his objectives because of the changes occurring in the world around him.

(My Day, April 16, 1946)

All the movements or crusades depend for their inception and early momentum on the personality of some individual man or woman. Someone must care tremendously and give unstintingly of themselves before an idea takes sufficient root.

(My Day, February 16, 1937)

One can have a bloodless revolution if one can count on leaders of sufficient vision to grasp the goal for which the mass of people is often unconsciously striving, and courage enough in the nation as a whole to accept the necessary changes to achieve the desired ends.

(*It's Up to the Women*)

Leaders are essential in a democracy, but it is the rank and file that make the progress.

("A Message to Parents and Teachers,"
Progressive Education, January–February 1934)

GOVERNMENT

The price of good government is constant vigilance and the participation of a strong and influential people who want good government on an active basis.

(My Day, November 8, 1945)

We, the people, today are in a period of retrogression. We do not want to be reminded of our unpleasant shortcomings, we do not want to face up to the big problems that we have to meet as a great people if we accept our place of leadership in the world. It is much easier and pleasanter to be a little people and so much less responsibility.

(My Day, May 18, 1946)

Our representatives in Congress are becoming temporizers. They are unable to make up their minds about anything because the American people are not clear and determined in their own minds where they stand and where they are going.

(My Day, May 7, 1946)

You may not care what your government does, but most of you do care whether anyone listens to you when you want a better school, some new housing or some new recreational facilities in your neighborhood. Unless you have voted for your local candidates, they won't listen to you on these subjects which touch your daily lives. It is the fact that you have a vote that makes you important to them.

(My Day, October 10, 1960)

I am inclined to think that if a question as serious as going to war were presented to our nation we would demand facts unvarnished by interpretation. Whether we even in our free democracy could obtain them is another question.

(*This Troubled World*)

Indifference, apathy, unwillingness on the part of good people to go down into the arena and fight, will give any city or any country poor government.

(My Day, November 8, 1945)

Wherever we find this growing tendency toward apathy, we ought to fight it tooth and nail. There could be no more destructive quality for America and its way of life.

(*Tomorrow Is Now*)

Governments do not become corrupt unless their citizens have allowed low standards to exist.

(If You Ask Me, *McCall's*, July 1951)

In the final analysis, a democratic government represents the sum total of the courage and the integrity of its individuals. It cannot be better than they are.

(*Tomorrow Is Now*)

Only with equal justice, equal opportunity, and equal participation in the government can we expect to be a united country.

("Social Gains and Defense,"
Common Sense, March 1941)

LAW

One thing no one can dispute: If you want a world ruled by law and not by force you must build up, from the very grassroots, a respect for law. It is the code we have created for our mutual safety and well-being. It is our bulwark against chaos. It is the fabric of our civilization.

(*You Learn by Living*)

While a law is in force it should be lived up to. You have a right to try to change it, and if you can persuade the majority that it should be changed then it will be changed. But you must abide by a law until it is changed or you will be flouting the fundamental things that make us a democracy and a law-abiding nation.

(My Day, January 4, 1956)

When you have put things on paper you haven't actually accomplished anything. The people have to accept changes, and when you are changing age-old customs this is sometimes difficult. . . . In a police state an edict may be enforced, but under other forms of government the people have to be persuaded and convinced and that takes time and education.

(My Day, May 25, 1951)

DEMOCRACY

Once more we are in a period of uncertainty, of danger. . . .
Once more we need the qualities that inspired the development of the democratic way of life. We need imagination
and integrity, courage and a high heart. We need to fan the
spark of conviction, which may again inspire the world as
we did with our new idea of the dignity and the worth of
free men.

(*Tomorrow Is Now*)

America is not a pile of goods, more luxury, more com-
forts, a better telephone system, a greater number of
cars. America is a dream of greater justice and opportunity for the average man and, if we can not obtain it, all
our other achievements amount to nothing.

(My Day, January 6, 1941)

It is opportunity which is the lifeblood of democracy.

(*Mrs. Eleanor Roosevelt's Own Program*, June 20, 1940)

The theory of a Democratic way of life is still a belief
that as individuals we live co-operatively, and, to the
best of our ability, serve the community in which we
live, and that our own success, to be real, must contribute to the success of others.

(*The Moral Basis of Democracy*)

Democracy believes in the right of people to develop peacefully and in the right of discussion and the rule of the majority. People may change their opinions, but they must do so under the rule of law and through persuasion and not force. . . ."

(Address at Stuttgart, Germany, October 23, 1948)

In a democracy we must be able to visualize the life of the whole nation.

("Insuring Democracy," *Collier's*, June 15, 1940)

One of the basic elements in our whole way of life, one of the elements that, for two hundred years, has attracted the peoples of the world to our shores, is our belief that each individual has a right to a better life and to better working conditions and a larger share of the prosperity of his country because, by having them, *he can develop his maximum potentialities as a human being.*

(*Tomorrow Is Now*)

Democracy cannot be static. Whatever is static is dead.

("Let Us Have Faith in Democracy,"
Land Policy Review, January 1942)

Sometimes the processes of democracy are slow, and I have known some of our leaders to say that a benevolent dictatorship would accomplish the ends desired in a much shorter time than it takes to go through the democratic processes of discussion and the slow formation of public opinion. But there is no way of insuring that a dictatorship will remain benevolent or that power once in the hands of a few will be returned to the people without struggle or revolution. This we have learned from experience and we accept the slow processes of democracy because we know that short-cuts compromise principles on which no compromise is possible.

("The Struggle for Human Rights,"
September 28, 1948)

Our solidarity and unity can never be a geographical unity or a racial unity. It must be a unity growing out of a common idea and a devotion to that idea.

(*The Moral Basis of Democracy*)

We must really believe in democracy and in our objectives. We cannot live in fear of either Fascism or Communism. We have to be certain that the majority of our people recognize the benefits of democracy and therefore are loyal to it.

(My Day, March 27, 1947)

Democracy requires both discipline and hard work. It is not easy for individuals to govern themselves. . . . It is one thing to gain freedom, but no one can give you the right to self-government. This you must earn for yourself by long discipline.

(*Tomorrow Is Now*)

If democracy is to succeed, we need well-disciplined citizens whose use their citizenship with intelligence.

(My Day, May 10, 1946)

This country can, and must, show that democracy isn't just a word, but that it means regards for the rights of human beings; that it means that every human being, regardless of race or creed or color, has equal dignity and equal rights; that it means that we care about the kind of freedom which allows people to grow, and allows them to develop their own potentialities and their own interests; that we recognize that democracy, as a basis for government, has to assume certain basic obligations to its citizens.

("What I Think of the United Nations,"
United Nations World Magazine, 1949)

Democracy will mean little to people of the world unless those who believe in it, work for it with unselfish and sacrificial spirit.

(My Day, July 4, 1946)

We in this country who believe in democracy and in a free-enterprise system will have to justify, not by lip service but by actual accomplishments, the claims which we make for our system of government and our political and economic way of life.

(My Day, April 3, 1946)

If we are honest with ourselves today, we all acknowledge that the ideal of Democracy has never failed, but that we haven't carried it out, and in our lack of faith we have debased the human being who must have a chance to live if Democracy is to be successful.

(*The Moral Basis of Democracy*)

We have neglected to remember that the rights of all people to some property are inviolate. We have allowed a situation to arise where many people are debased by poverty or the accident of race, in our own country, and therefore have no stake in Democracy; while others appeal to this old rule of the sacredness of property rights to retain in the hands of a limited number the fruits of the labor of many.

(*The Moral Basis of Democracy*)

We accepted our freedom as a gift from the pioneers and from heaven, and yet it is more than evident today that there are constant assaults on our liberty, perhaps not the least of which is our own apathy. If we wish democracy to survive we must be constantly alive to the many-sided battle we wage.

(If You Ask Me, *Ladies' Home Journal*, May 1941)

Some people feel that human nature cannot be changed, but I think when we look at what has been achieved by the Nazi and Fascist dictators we have to acknowledge the fact that we do not live in a static condition, but that the influences of education, of moral and physical training have an effect upon our whole beings. If human beings can be changed to fit a Nazi or Fascist pattern or a Communist pattern, certainly we should

not lose heart at the thought of changing human nature
to fit a Democratic way of life.

(*The Moral Basis of Democracy*)

We as a democracy in these times must be able to grasp
our problems, must have sufficient general education
to know not only what our difficulties are but what the
government is trying to do to help us meet those difficulties. Without that ability in our people and without
the willingness to sacrifice on the part of the people as
a whole, in order that the younger generation may develop this ability, I think we have harder times ahead.

("Negro Education," speech to the National Conference
on the Education of Negroes, May 11, 1934)

The most powerful weapon that we have at our com-
mand to-day is public opinion. Statesmen quail before
it and it could move mountains!

(My Day, February 28, 1936)

If we cannot prove here that we believe in freedom, in
the ability of people to govern themselves, if we cannot have confidence in each other, if we cannot feel that
fundamentally we are all trying to retain the best in our
democracy and improve it, if we cannot find some basic
unity even in peace, then we are never going to be able
to lead the world.

(Address to Americans for Democratic Action,
April 1, 1950)

TOTALITARIANISM

If people do not understand what problems they face, it leads to apathy, to indifference, to shifting the responsibility that each individual ought to feel. That must be avoided if a democracy is really going to be a democracy.

("In Service of Truth," *The Nation*, July 9, 1955)

Among free men the end cannot justify the means. We know the patterns of totalitarianism—the single political party, the control of schools, press, radio, the arts, the sciences, and the church to support autocratic authority; these are the age-old patterns against which men have struggled for three thousand years. These are the signs of reaction, retreat, and retrogression.

("The Struggle for Human Rights," September 28, 1948)

I believe in defending democracy by preserving democracy, not by descending to the use of the same weapons used in dictatorships.

(My Day, May 5, 1948)

It is easier at first to follow a fascist because he promises more and accepts greater responsibility, leaving the average human being, who dislikes responsibility, feeling that more will be done for him and that he himself need make no exertion.

(*If You Ask Me*, 1946)

In the long run, even a totalitarian government must listen to the wishes of its people.

(My Day, July 12, 1957)

Unless we learn to live together as individuals and as groups, and to find ways of settling our difficulties without showing fear of each other and resorting to force, we cannot hope to see our democracy successful. It is an indisputable fact that democracy cannot survive where force and not law is the ultimate court of appeal.

("Keepers of Democracy,"
Virginia Quarterly Review, January 1939)

CITIZENSHIP

I do believe that every citizen, as long as he is alive and able to work, has an obligation to work on public questions and that he should choose the kind of work he is best fitted to do.

("Why I Do Not Choose to Run," *Look*, July 9, 1946)

The chief duty of the citizen is to make his government the best possible medium for the peaceful and prosperous conduct of life.

(*You Learn by Living*)

Because it is the most highly developed type of government, democracy requires the most highly developed citizens. Whether we, as a whole, are willing to accept this responsibility and live up to it is something of which I cannot be sure. But we cannot be reminded too often that each of us is responsible for our attitude and our way of life, because they will in turn affect our government.

(*You Learn by Living*)

Each individual in a democracy has the duty to live up to the standards which he believes are right. This may sometimes be a disagreeable duty. But when you believe in the right of equality among men, then you must accept the duty of individual thinking and living by your own standards.

(My Day, May 6, 1946)

As individuals we feel that our action or inaction will produce about the same result—nobody in authority will pay attention to what we think! Until we free ourselves of this inferiority complex which is nothing more than a comfortable alibi to sidestep responsibility, I do not see much chance of improving conditions either at home or abroad.

(My Day, February 28, 1936)

Telegrams and communications sent to Washington have very little effect unless they show that the individual really knows what he thinks and intends to awaken people in his community. That, I think, is the role which every individual citizen can play. All of us have a small orbit of influence where, if we have convictions, we can stand up for them and even persuade others to join with us.

(*If You Ask Me,* 1946)

Knowing our own country is something which, while we can consider it recreation, is closely allied to a duty, for it is necessary for good citizenship that we should attempt to visualize the vastness of our country and the multiplicity of interests amongst our people.

(*It's Up to the Women*)

Confidence in your fellow citizens can only be maintained in a civilization which makes certain standards of living possible for each and every individual and gives an equal opportunity to every individual to progress as far as his ability and character will permit.

(*It's Up to the Women*)

Our people want to prove that men are capable of governing themselves; that they are capable of learning to live together at home in peace, and therefore to live peacefully with the rest of the world. This means an individual soul-searching and discipline which no people in any nation has ever undertaken before.

(My Day, March 6, 1950)

Though we may occasionally find among us people who will lie, steal, cheat and murder, we must write them off as the failures resulting from a civilization which, while it is better, I believe, than that of any other country, is still not perfect. The great mass of our people, however, are capable of greatness and complete adequacy, of fine dreams, of courage often surpassing that of their leaders.

(My Day, January 2, 1951)

I can't help believing that most of us want to be good citizens and that we will do all we can to fulfill our obligations once we clearly understand them.

(My Day, May 21, 1946)

It takes great determination to go on working, year after disappointed and frustrating year, for some reform that seems important to you. As time passes you feel that nothing has been accomplished. *But, if you give up, you are abandoning your own principles.* It is deeply important that you develop the quality of stamina; without it you are beaten; with it, you may wring victory out of countless defeats, after years of what seemed to be hopeless effort.

(*Tomorrow Is Now*)

PATRIOTISM

Everyone of us cares more for his own country than for any other. That is human nature. We love the bit of

land where we have grown to maturity and known the
joys and sorrows of life.

(My Day, December 22, 1945)

Real patriotism means that you look ahead to what is
going to benefit your own country—and today it is going
to benefit our own country to learn to work with the
other nations of the world.

("Why the United Nations Is Unpopular—
And What We Can Do About It," November 19, 1952)

The flag symbolizes for me the nation which we all love.
When we pledge allegiance to the flag we pledge allegiance not only to our country but to the freedoms and
principles which have built this nation.

(If You Ask Me, *Ladies' Home Journal,* January 1954)

VOTING

The minimum, the very basic minimum, of a citizen's
duty is to cast a vote on election day. Even now, too few
of us discharge this minimal duty.

(*You Learn by Living*)

An individual may not like the decision of the majority,
but if he overthrows it he has destroyed his own right

to change his mind. The thing that really preserves freedom is the liberty of every individual to express his thinking of the moment; and, since thinking can change, individuals in the minority need never feel they are not being fairly treated. They have every opportunity through education to change that minority into a majority.

("Let Us Have Faith in Democracy,"
Land Policy Review, January 1942)

A democratic form of government is dependent for its success on an informed voter. It is only easy to over throw democracy where the sources of information are not available to the average citizen.

(My Day, August 21, 1936)

A vote is never an intelligent vote when it is cast without knowledge.

(*It's Up to the Women*)

The right to vote should be something every citizen of this country enjoys without any question.

(My Day, April 11, 1960)

It seems to me that one should say to all citizens, no matter of what race, color or national origin, that their minimum duty is to use the right of participation by vote in their government. There is an obligation to make the vote meaningful by a study of the issues and personalities, but not to use it is a denial of one's belief in democracy and weakens the structure of our society.

(My Day, October 2, 1961)

I certainly do not think our failure to vote is part of our democratic system. I think it stems from the fact so many people believe that all rights and freedoms that go with democracy are going on regardless of what we do. It is not made clear to us from the time we are little children, at home and in the schools, that our duties as citizens in a democracy come before any other duties. . . ."

(If You Ask Me, *McCall's,* November 1950)

Many people do not vote because they honestly do not understand what the policies back of the oratory are, and they decide that it will not matter much which party wins and therefore they stay away from the polls.

("If I Were a Republican Today," *Cosmopolitan,* June 1950)

Two

RACE AND ETHNICITY

IN AN ERA WHEN RACIAL DISCRIMINATION AND anti-Semitism were commonplace, Eleanor Roosevelt dared to make equality a central theme of her activism. She did so because she believed America had a moral and political obligation to live up to the promise of its founders: "Unless we learn to live together as individuals and as groups, and to find ways of settling our difficulties without showing fear of each other and resorting to force, we cannot hope to see our democracy successful," she wrote in 1939.[1]

In her mind, a successful democracy meant that freedom, justice, and equality of opportunity were available to all citizens no matter their race, ethnicity, or religious background. "The problem is not to learn tolerance . . . ," she wrote, "but to see that all alike have hope and opportunity."[2]

While she thought those goals were achievable, she never minimized the difficulty of reaching them. She understood that much of the prejudice directed toward African Americans, immigrants, refugees, and other minorities was rooted in fear and ignorance. Changing those attitudes required time and effort on the part of individuals and communities. She continually challenged her fellow Americans to examine their own attitudes, work through them, and then come together in support of a more just and inclusive society.

Failure to change entrenched exclusionary attitudes, she warned, would have implications far beyond America's borders. "If we cannot meet the challenge of fairness to our citizens of every nationality, of really believing in the Bill of Rights and making it a reality for all loyal American citizens, regardless of race, creed, or color; if we cannot keep in check antisemitism, anti-racial feelings as well as anti-religious feelings, then we shall have removed from the world, the one real hope for the future on which all humanity must now rely."[3]

Eleanor's stance won her few friends. She was often vilified and accused of stoking racial discord. She even received death threats. Nevertheless, she persisted in her beliefs and acted on her conviction that what was important was "not what religion or race we belong to, but how we live our lives."[4]

TOLERANCE

For the people of this country, the question is whether they can continue to exist without giving all citizens equality before the law and equal dignity as human beings. We must make this decision and upon it depends our whole future and that of white peoples everywhere.

(My Day, May 5, 1956)

We have no common race in this country, but we have an ideal to which all of us are loyal. . . . It is an ideal which can grow with our people, but we cannot progress if we look down upon any group of people among us because of race or religion. Every citizen in this country has a right to our basic freedoms, to justice and equality of opportunity, and we retain the right to lead our individual lives as we please, but we can only do so if we grant to others the freedoms that we wish for ourselves.

("A Challenge to American Sportsmanship," *Collier's*, October 16, 1943)

What is really important is not what religion or race we belong to, but how we live our lives.

(My Day, June 19, 1943)

I am afraid with a good many people tolerance is a matter of indifference. But, when it has its roots in the security of one's convictions and beliefs, then tolerance can be a very fine thing. In that kind of tolerance there is true humility which, in spite of personal conviction, listens and tries to understand other points of view.

(My Day, May 26, 1942)

Real tolerance does not attempt to make other people conform to any particular religious or racial pattern.

("Intolerance," *Cosmopolitan*,
February 1940)

The time has come when the fight must be made by each one of us to live at home in a way which will make it possible to live peacefully in the world as a whole.

(My Day, February 9, 1948)

We must show by our behavior that we believe in equality and in justice and that our religion teaches faith and love and charity to our fellow men.

(*The Autobiography of Eleanor Roosevelt*)

The day of selfishness is over; the day of really working together has come, and we must learn to work together, all of us, regardless of race or creed or color; we must wipe out, wherever we find it, any feeling that grows up of intolerance, of belief that any one group can go ahead alone. We go ahead together or we go down together.

("Negro Education," speech to the National Conference on the Education of Negroes, May 11, 1934)

We must keep moving forward steadily, removing restrictions which have no sense, and fighting prejudice. If we are wise we will do this where it is easiest to do it first, and watch it spread gradually to places where the old prejudices are slow to disappear.

("Race, Religion and Prejudice," *New Republic*, May 11, 1942)

Tolerance ought only to be the preliminary step which allows us to get to know other people, and which prevents us from setting up bars, just because they may be of a different race or religion. The real value of any relationship is the fact that we learn to like people in spite of our differences.

(My Day, November 24, 1943)

Mutual respect is the basis of all civilized human relationships.

(*You Learn by Living*)

INTOLERANCE

It would seem that we persist in doing all that we can to stir up the very forces around us, which we profess to want to allay. Instead of acting with kindness, we seem to do the very things which promote intolerance and hatred amongst races and religious groups, to say nothing of the way we treat each other when we happen to be labelled workers or employers.

(My Day, July 31, 1939)

All intolerance is based on fear, and fear is usually a lack of understanding.

("Defense and Girls," If You Ask Me,
Ladies' Home Journal, May 1941)

When people are made uncomfortable, the spirit of tolerance disappears. No longer are they interested in what is right or wrong; they want action and a return to comfortable living conditions.

("Intolerance," Cosmopolitan, February 1940)

People who are having a hard time will complain about certain races, or certain groups, because they fear the economic competition of certain individuals, who either because of racial background or because of former experiences, may succeed better than they have succeeded.

("Intolerance," Cosmopolitan,
February 1940)

I do not believe that oppression anywhere or injustice which is tolerated by the people of any country toward any group in that country is a healthy influence.

("Keepers of Democracy,"
Virginia Quarterly Review, January 1939)

RACIAL AND RELIGIOUS PREJUDICE

Prejudice so blinds us that we see only what we expect to see, what we want to see.

(*Tomorrow Is Now*)

We are to blame for much of the bigotry, ignorance and vice in this country because so few of us think it necessary to do more than keep quiet.

(My Day, May 29, 1936)

We never know where prejudices will lead us. Neither do we know how often we use our prejudices to excuse or cloak motives and emotions which we would be ashamed to bring into the light of day. . . .

(My Day, December 16, 1944)

When you begin to allow yourself to override the law you do not know where you will end. When you begin to allow yourself a kind of self-righteous prejudice against another race or another religion, you do not know what the end may be, and in the end you may suffer. . . .

(My Day, April 29, 1959)

If we cannot meet the challenge of fairness to our citizens of every nationality, of really believing in the Bill of Rights and making it a reality for all loyal American citizens, regardless of race, creed or color; if we cannot keep in check antisemitism, anti-racial feelings as well as anti-religious feelings, then we shall have removed from the world, the one real hope for the future on which all humanity must now rely.

(My Day, December 16, 1941)

When we permit religious prejudice to gain headway in our midst, when we allow one group of people to look down upon another, then we may for a short time bring hardship on some particular group of people, but the real hardship and the real wrong is done to democracy and to our nation as a whole. We are then breeding people who cannot live under a democratic form of government but must be controlled by force.

("Keepers of Democracy,"
Virginia Quarterly Review, January 1939)

You can have no real democracy when the people in your midst, whatever their race or color or creed, and whatever their crime may be, cannot be sure of a fair trial and even-handed justice. No one wants sentimental kindness, but all men under whatever government they live want freedom and justice. There is no freedom when one group of individuals can strike fear into the hearts of other individuals and use violence against them.

(My Day, July 29, 1946)

The psychology which believes that the white man alone of all the races in the world, has something which must be imposed on all other races, must go. We know today that our chance to live in peace in the future lies in respect for the individual, no matter what his color. We must have a willingness to accept what anyone has to contribute to civilization, and to cooperate in the difficult business of "live and let live."

(My Day, April 3, 1942)

No one can honestly claim that either the Indians or the Negroes of this country are free. These are obvious examples of conditions which are not compatible with the theory of Democracy. We have poverty which enslaves, and racial prejudice which does the same.

(*The Moral Basis of Democracy*)

Had we done a really good job it seems to me that our Indians today would be educated; there would be no need of reservations; they would be fully capable of taking their places as citizens, and the tribes would have

had full compensation for the lands they owned. Our
inability to work out this small problem satisfactorily
and fairly is one of the real blots on our history.

(My Day, August 25, 1950)

Until we have complete equality of opportunity in every
field, equal rights socially and economically, we cannot
consider ourselves a real democracy.

(My Day, December 11, 1945)

Segregation has not always been based on color. Sometimes it has been based on religion or sex. It can take on
many forms and one finds it between people of the same
color, simply based on nationality or caste. But segregation in all its strange patterns is on its way out because
of the wave of freedom which is passing over the world.

("Segregation," *Educational Forum*,
November 1959)

There cannot be, of course, complete equality for every
human being because, even though we have equal opportunity, our native gifts and the circumstances in
which we are born condition our development. But
our race and our religion should not place any special
handicaps upon us. That is the concept on which these
United States came into being, and the sooner we bring
it to fulfillment, the sooner will the dreams of many of
our people come true.

(My Day, December 11, 1945)

Discrimination of any kind leaves scars on the human soul.

(My Day, December 2, 1960)

IMMIGRANTS

I wonder how many of us would be here today if the founding fathers had been as nervous as we are about the oncoming hordes that threatened to starve them to death when they were not growing much more than they themselves could eat!

(My Day, November 20, 1946)

So much that is good has come to us through our foreign immigration, and yet so often we hear people who seem to be entirely ignorant of the contribution which other lands have made to civilization in the United States. We would undoubtedly lose much if we ever cut off entirely this flow of new blood into our country.

I think there is much to be said, however, in favor of deporting alien criminals who are not citizens. Secondly, I would like to see all people who live here and earn their living here, make up their minds after a given period of time either to go home, or to become American citizens. The exact period when they should be asked to make this final decision is, of course, debatable, but it should be at some point and not too long after they have had an opportunity of knowing this country. This seems to me entirely fair and desirable for the good of the nation.

(My Day, May 24, 1939)

Many a good citizen not born in this country has, perhaps, a greater appreciation of what it means to live in a free and self-governing country and, therefore, a greater sense of responsibility for preserving the democratic way of life.

(My Day, August 7, 1943)

In this country we have meant to be the great melting pot of the world. . . . The variety of our backgrounds has been one of our great strengths. We owe our continued vigor over a long period of years to the adventurous blood which has been pumped into our veins. People who stay at home, too dispirited to get out and seek something better than they have known before, do not come to new worlds.

("From the Melting Pot—An American Race," *Liberty*, July 14, 1945)

Not a little of our success in the past has come because we have had new people coming to our country with fresh ideas and fresh determination to succeed.

(My Day, June 30, 1947)

We built this country on the labor of immigrants and on the humanitarian principles that the Statue of Liberty personifies. We said we were a haven for the oppressed of the world. We can no longer open our doors as we did in the early days because ours is now a highly developed nation, but we are still able to preserve some of our humanitarianism and to profit by the skills and the strength of a certain amount of immigration. It would be wrong, I think, to say that we should take no one into our country from now on.

(My Day, May 8, 1953)

We must be a "land of the free" or too many people will cease to believe that there is a place anywhere in the world where freedom and equality of opportunity exist for human beings.

(My Day, December 13, 1947)

REFUGEES

I think many of us have forgotten that our ancestors came here and made this a land of refuge, and that some of our best citizens have been the sons and daughters of refugees!

(My Day, March 5, 1946)

If we study our own history we find that we have always been ready to receive the unfortunates from other countries, and though this may seem a generous gesture on our part, we have profited a thousand fold by what they have brought us.

(My Day, January 23, 1939)

In the past, America has taken in refugees from many lands. We have learned the way to help: many people in many communities, through their church organizations and other groups, know the techniques of assisting refugees to adjust in a new community.

(My Day, April 2, 1962)

The refugees of the world are a constant and painful reminder of the breakdown of civilization through the stupidity of war. They are its permanent victims.

(*The Autobiography of Eleanor Roosevelt*)

A new type of political refugee is appearing—people who have been against the present governments and if they stay at home or go home will probably be killed.

(Diary entry, London, January 9, 1946)

I know well that there are people in this country who, for a number of very obvious reasons are afraid to bring anyone in from the displaced persons' camp. I think this fear is far more dangerous than the people whom we might bring in. We are growing so pusillanimous that, in a short time, we will be afraid of our own shadows—and this is no world in which the fearful survive. Good new blood will do us no harm. We are suffering from fear of the unknown. But these people have seen so much that was unknown that they have learned how to accept it and stand up to it.

Here I am, blaming Congress when, after all, the blame really lies with ourselves. It is we who have not been articulate enough and have not educated people as a whole to understand that the displaced persons are just like ourselves. Only, for the moment, they have no country. We have not made people realize to the full what the ruin of Europe will mean in lowering our own standards of living. Again we the people have failed—failed ourselves and failed our brethren overseas.

(My Day, July 23, 1947)

Three

FREEDOM AND RIGHTS

THE GREAT DEPRESSION, WORLD WAR II, AND THE COLD War severely strained democratic institutions, and the pressure to restrict civil liberties, especially during the Second World War and its aftermath, was intense.

Eleanor Roosevelt recognized the danger. Freedom was a precious thing, easily lost unless it was vigilantly protected. "Men who have the instincts for dictatorship are always a danger in any society," she once observed. "Free citizens must be constantly alert to preserve their liberties."[1] In the middle of World War II she wrote, "At all times, day by day, we have to continue fighting for freedom of religion, freedom of speech, freedom from fear, and freedom from want—for these are things that must be gained in peace as well as in war."[2]

At the same time, she recognized that freedom in a

democracy is contingent upon the freedom of one's fellow citizens, arguing that "we retain the right to lead our individual lives as we please, but we can only do so if we grant to others the freedoms that we wish for ourselves."[3]

She opposed those like Representative Martin Dies Jr. (D-TX), chair of the House Un-American Activities Committee (HUAC), and Senator Joseph McCarthy (R-WI), whose views on the impact of communism in American life she considered harmful. Though staunchly anticommunist herself, she deplored the smear tactics they and their colleagues used during the Red Scare in the late 1940s and early 1950s to stifle political dissent, saying they resembled the methods the Soviets used to control people's minds.[4]

In an era when fear and paranoia dominated public discourse, her principled stand was far from popular. Even so, she consistently defended civil liberties, arguing that "the test of democracy and civilization is to treat with fairness the individual's right to self-expression even when you can neither understand nor approve it." To her, stifling dissent in order to combat suspected communist incursion was a betrayal of the founders' vision for America. Instead she challenged her fellow citizens to overcome their fears, reclaim their freedoms, and concentrate on the fundamental goal of making democracy work for everyone.[5]

FREEDOM

Only in freedom can a man function completely.
(*The Autobiography of Eleanor Roosevelt*)

I'm intensely anxious to preserve the freedom that gives
you the right to think and to act and to talk as you please.
That I think is essential to happiness and the life of the
people.
(Interview with Mike Wallace, November 23, 1957)

Freedom seems to be a whole. And if you want to have
it at home I think you have to fight for it too abroad.
You have to hope for it everywhere because you cannot
preserve it just in your own country. It will not stay in
one place. . . ."
(*The Fears of Free Americans*, student council pamphlet,
March 26–28, 1954)

Freedom for our peoples is not only a right, but also a
tool. Freedom of speech, freedom of the press, freedom
of information, freedom of assembly—these are not just
abstract ideals to us; they are tools with which we create a way of life, a way of life in which we can enjoy
freedom.
("The Struggle for Human Rights,"
September 28, 1948)

We must not be confused about what freedom is. Basic human rights are simple and easily understood: freedom of speech and a free press; freedom of religion and worship; freedom of assembly and the right of petition; the right of men to be secure in their homes and free from unreasonable search and seizure and from arbitrary arrest and punishment.

We must not be deluded by the efforts of the forces of reaction to prostitute the great words of our free tradition and thereby to confuse the struggle. Democracy, freedom, human rights have come to have a definite meaning to the people of the world which we must not allow any nation to so change that they are made synonymous with suppression and dictatorship.

("The Struggle for Human Rights,"
September 28, 1948)

The American way of life means to me freedom to hear all sides of a question; to state my opinion even where the question concerns my government and its officials.

It means to me the right of association with people I desire to join with for work or for pleasure.

It means belief in civil liberties and an effort to see that they are equal for all people within my country and that opportunity is open to all on an equal basis.

(Letter to Elsa Marcussen,
February 2, 1948)

It seems to me that if you curtail freedom in any respect, you curtail it sooner or later in every respect.

(Remarks at the meeting of the UN General Assembly
Third Committee, October 28, 1947)

The price paid for the results obtained under all forms of totalitarian government is the surrender of individual freedom.

(My Day, June 1, 1948)

It is my deep conviction that any society which does not provide freedom for the upcoming generations to work openly and honestly for their aspirations contains within it the seeds of its own destruction.

(Address to Les Jeune Amis de la Liberté,
December 18, 1951)

Great power is always dangerous to freedom and unless we share this power with the other nations of the world, and are ever careful to make it of value to the people of the world, we may find ourselves becoming less the defenders of freedom than the custodians of power.

(My Day, May 30, 1944)

The kind of freedom which we desire is the kind that leaves one free to search for truth, to discover facts wherever possible, to experiment freely in the fields of science, to give to creative artists in music, in drama and writing, in painting or sculpture, the chance to express themselves uncensored by the government and unencumbered by any political dogma. . . .

We recognize, of course, that all those who enjoy freedom must learn self-discipline—not discipline imposed by the state but discipline imposed by themselves for the sake of the rights of other human beings.

(Statement for Paris, April 14, 1949)

Freedom makes a huge requirement of every human being. With freedom comes responsibility. For the person who is unwilling to grow up, the person who does not want to carry his own weight, this is a frightening prospect.

(*You Learn by Living*)

Freedom must always be qualified by the fact that your own freedom must not mean somebody else's slavery.

(*It's Up to the Women*)

When you come to understand self-discipline you begin to understand the limits of freedom. You grasp the fact that freedom is never absolute, that it must always be contained within the framework of other people's freedom.

(*You Learn by Living*)

It is obvious that freedom is always conditioned by the amount of ability an individual has to govern himself.

(*If You Ask Me*, 1946)

Each of us has to make sure that freedom really exists, that in our hearts we have the spirit that really wants freedom not just for ourselves but for the other fellow too. We must want the kind of freedom in which people discipline themselves sufficiently so that everyone has rights.

("Suspicion as Peace Bar Feared by Mrs. Roosevelt," *Newark Evening News*, October 2, 1945)

At all times, day by day, we have to continue fighting for freedom of religion, freedom of speech, freedom from fear, and freedom from want—for these are the things that must be gained in peace as well as in war.

(My Day, April 15, 1943)

Above all, we should not allow our fears to develop the kind of security that is enforced in a police state and is, therefore, never real democratic security.

(My Day, September 28, 1949)

Today we are faced with a world in which all people are neighbors, and if we here become closed in and narrow and lose the freedoms that are traditional with us then I think we are going to do that same thing to the whole world and instead of leading to freedom we are going to lead to slavery.

(*The Fears of Free Americans*, student council pamphlet, March 26–28, 1954)

Above all other things, we must be jealous to preserve
the liberties that are inherent in true democracy.

(My Day, September 3, 1949)

CIVIL LIBERTIES

Men who have the instincts for dictatorship are always
a danger in any society. Free citizens must be constantly alert to preserve their liberties. In the United
States, it is easy to discover a demagogue, but it sometimes requires courage to stand up immediately and say
you don't agree with certain methods and certain ideas.
However, if we want to preserve our liberties, we had
better show that courage—it is the only way I know to
remain a free people.

(If You Ask Me, *McCall's*, November 1960)

When the power becomes concentrated in the hands of
a few, there is great danger that the majority will not be
able to move at all.

("Fear Is the Enemy," *The Nation*,
February 10, 1940)

The reason we have grown, and are growing, in this
country is because our minds have remained free.
We have always questioned our own wisdom; we have

been ready to make experiments and to change our mind.

(My Day, May 17, 1950)

We do not move forward by curtailing people's liberty because we are afraid of what they may do or say. We move forward by assuring to all people protection in the basic liberties under a democratic form of government, and then making sure that our government serves the real needs of the people.

("Fear Is the Enemy," *The Nation*, February 10, 1940)

To be content with what is being done—and not to examine whether something better can be done—may lead to less trouble for the government, but it does not lead to more accomplishments in the area of civil liberties.

(My Day, October 19, 1960)

This country is no more sure of preserving civil liberties, in spite of guaranties provided by law, than any other country if the people are not constantly alert. . . .

(My Day, May 15, 1950)

My own feeling is that when a people fight for something they are stronger than when they are merely trying to repress those who happen to think differently from themselves.

(My Day, September 28, 1949)

I have no patience with people who do not want to preserve our fundamental rights and freedoms and who will not live up to the Constitution.

(My Day, August 20, 1960)

FREEDOM OF SPEECH

Freedom is not really freedom unless you can differ in thought and in expression of your thought.

("Reply to Attacks on U.S. Attitude toward Human Rights Covenant,"
Department of State Bulletin, January 14, 1952)

All must have a right to be heard. When we begin to discriminate we can never tell where the discrimination may end.

(My Day, March 21, 1950)

Surely the public is not afraid of listening to viewpoints with which they do not agree.

(My Day, March 21, 1950)

Even if we do not agree with the political beliefs held by some, we must not reach a state of fear and hysteria which will make us all cowards! Either we are strong enough to live as a free people or we will become a police state. There is no such thing as being a bystander on these questions!

(My Day, August 13, 1947)

It is surely a terrible thing when young people think that what you think at eighteen means you can never think any differently, or that you are going to be held to account for what you thought at eighteen. It seems to me that means a loss of freedom.

("In Service of Truth," *The Nation,* July 9, 1955)

We have made it possible for people of opposing ideas to live without fear. We feel that a system which defends the right of the individual to speak his mind, even against his own government if necessary, is a safeguard to all freedom.

(My Day, August 2, 1946)

In the democracies of the world, the passion for freedom of speech and thought is always accentuated where there is an effort anywhere to keep ideas away from people and to prevent them from making their own decisions.

(My Day, May 11, 1943)

DEMAGOGUERY/PROPAGANDA

The forces against freedom understand that their own hope of imposing dictatorial regimes on new areas of the world lives in the disunity of the free world. Hence, they use every propaganda trick to sow confusion and dissension in the ranks of free peoples. They exploit every feeling of fear and antagonism to divide free nations, and break the spirit for collective resistance to aggression.

If we are determined not to lose our freedoms, we must use our heads in an active campaign to expose the propaganda designed to divide us, and to promote the unity and cooperation of free people.

(Address to Les Jeune Amis de la Liberté,
December 18, 1951)

When you openly now attack not the way some high official is conducting the business of the nation, or the business for which he is responsible, but his personal character and his intentions, then I think you are giving the world today a proof of the fact that you don't trust your officials to live up to loyalty to their country, to the oath which they took to support their country, and that you as a people in a democracy have chosen to be your leaders people of such questionable integrity and ability that you are criticizing not the way that they are carrying out certain policies but actually their character and their intentions and their human qualities which are important in making people respect the office and the government.

(*The Eleanor Roosevelt Program*, August 23, 1951)

This recurring matter of labeling "Communist" anyone who does not agree with you is essentially an act of dishonesty and it should be nailed every time for what it is.

(Tomorrow Is Now)

We should realize that the truth about complex problems is harder to understand than slogans and emotional appeals that do not meet the issue. There-fore, those who wish to defend their freedoms have a difficult task of education to perform constantly in order to prevent the sloganized propaganda from mis-leading people.

(Address to Les Jeune Amis de la Liberté,
December 18, 1951)

A tyrant can never tell who is for him or against him because he cannot enter the secret heart of any man.

(Address to Les Jeune Amis de la Liberté,
December 18, 1951)

FREEDOM OF ASSEMBLY

Guilt by association is a very dangerous accusation and the smearing of people before you have proved them guilty puts our country in a strange light before the rest of the world.

(My Day, March 22, 1950)

I think if we care for the preservation of our liberties we must allow all people whether we disagree with them or not, to hold meetings and express their views unmolested as long as they do not advocate the overthrow of our Government by force.

(My Day, September 3, 1949)

FREEDOM OF INFORMATION

We in the United States have become so intoxicated by our new methods of communication that we have failed to look closely at just what we are communicating—or failing to communicate.

(*Tomorrow Is Now*)

One of the best ways of enslaving a people is to keep them from education and thus make it impossible for them to understand what is going on in the world as a whole. . . .

The second way of enslaving a people is to suppress the sources of information, not only by burning books but by controlling all the other ways in which ideas are transmitted.

(My Day, May 11, 1943)

While accepting the fact that some of our press, our radio commentators, our prominent citizens and our movies may at times be blamed legitimately for things they have said and done, still I feel that the fundamental right of freedom of thought and expression is essential. If you curtail what the other fellow says and does, you curtail what you yourself may say and do.

(My Day, October 29, 1947)

I have stated that although I frequently disagreed with the opinions expressed by certain groups of papers in this country, I would hesitate to take any steps to curtail their freedom of expression. This, because when you begin to prevent the expression of opinion because you do not like it, you may shortly find that you curtail the expression of opinion which you like. Freedom of expression must be guaranteed to all, not just to those who think in one particular way.

(My Day, June 25, 1951)

I am afraid if we started trying to control the press, we might really do away with an essential freedom. It is true that this freedom is often abused, but I think the basis of democracy is that we educate people sufficiently well so that they can be trusted, in the long run, to judge political propaganda and the type of news that is slanted by certain types of interests. The individual citizen can best control the press by insisting always that the papers in his own home environment write uncolored news stories.

(If You Ask Me, *Ladies' Home Journal*, March 1947)

A really free press is a very difficult thing to obtain, and I often wonder whether those who clamor loudly for it in this country really want a free press or a press which they are free to run in their own way.

(My Day, March 7, 1947)

The check of a really free press is valuable not only over politicians, but over capital and labor as well.

(My Day, October 23, 1942)

No one believes in printing untruthful news. Certainly no one wants to see the press or individuals or any organizations incite to war. But I cannot believe suppression by law or by government edict is going to bring about the desired results.

(My Day, October 27, 1947)

I think the only hope for a really free press is for the public to recognize that the press *should* not express the point of view of the owners and the writers but be factual, whereas editorials *must* express the opinions of the owners and writers. . . . If [the public] really want to get at the truth, they can read a variety of publications whose owners and writers have different points of view and in so doing they will be able to decide where they themselves stand.

If owners and writers express what they honestly believe and are not influenced by advertisers or investors, we will have as nearly an honest press as it is possible to obtain.

(If You Ask Me, 1946)

The harm that can be done by censorship seems to me greater than the harm that could be done by a few people who do not have that high sense of responsibility and integrity.

(My Day, June 9, 1951)

RELIGIOUS FREEDOM

Religious freedom cannot just be Protestant freedom. It must be freedom for all religions.

(My Day, September 20, 1960)

Like all our other freedoms, this freedom from religious-group pressure must be constantly defended.

(*The Autobiography of Eleanor Roosevelt*)

Our Democracy in this country had its roots in religious belief, and we had to acknowledge soon after its birth that differences in religious belief are inherent in the spirit of true Democracy. Just because so many beliefs flourished side by side, we were forced to accept the fact that "a belief" was important, but "what belief" was important only to the individual concerned.

(*The Moral Basis of Democracy*)

It seems to me that the thing we must fix in our minds is that from the beginning this country was founded on the right of all people to worship God as they saw fit, and if they did not wish to worship they are not forced to worship. That is a fundamental liberty.

("Civil Liberties—The Individual and the Community," address to the Chicago Civil Liberties Committee, March 14, 1940)

For just as the Soviet Union has made a religion out of a political creed—communism—so, by a kind of reverse twist, the followers of Islam have made their religion an integral and controlling part of their political life.

(*India and the Awakening East*)

I would like to make it clear once and for all that I believe in the right of any human being to worship God according to his conviction, and I would not want to see this right taken away from anyone.

(My Day, July 8, 1949)

The freedom to belong to any church or to none is fundamental, however, to our rights as citizens, and it should not be used to prevent us from holding public office, which is also a fundamental right of every citizen.

(My Day, October 14, 1960)

I think the only real danger of our curtailing religious rights lies in the possibility that some of our church groups might come to wield too much influence in the nation's political and economic life. I think that would provoke very serious opposition because of the strong feeling in this country that the church should confine itself to spiritual matters, leaving affairs of government and the economy entirely free from church influence or domination.

("From the Melting Pot—An American Race,"
Liberty, July 14, 1945)

SEPARATION OF CHURCH AND STATE

The separation of church and state is extremely important to any of us who hold to the original traditions of our nation. To change these traditions by changing our traditional attitude toward public education would be harmful . . . to our whole attitude of tolerance in the religious area.

(My Day, June 23, 1949)

The people who settled in New England came here for religious freedom, but religious freedom to them meant freedom only for their kind of religion. They were not going to be any more liberal to others who differed with them in this new country, than others had been with them in the countries from which they came. This attitude seems to be our attitude in many situations today.

(*This Troubled World*)

We may belong to any religion or to none, but we must acknowledge that the life of Christ was based on principles which are necessary to the development of a Democratic state.

(*The Moral Basis of Democracy*)

The recognition of any church as a temporal power puts that church in a different position from any of the other churches and while we are now only hearing from the Protestant groups, the Moslems may one day wake up to this and make an equal howl. For us who take a firm stand on the separation of church and state, the recognition of a temporal power seems inconsistent.

(Letter to Harry S. Truman, January 29, 1952)

Spiritual leadership should remain spiritual leadership and the temporal power should not become too important in any church.

(Letter to Francis Cardinal Spellman, July 23, 1949)

Four

ECONOMICS

BOTH THE GREAT DEPRESSION AND THE RISING prosperity of the late 1940s and 1950s colored Eleanor Roosevelt's view of economics. She based her philosophy on what she called the "implicit" idea that "every man has a right to work."[1] Although she supported the free enterprise system, she did not believe that markets alone could ensure a fair and equitable economy. She thought government had a role to play in preventing "the strong from removing all opportunities from the weak."[2]

She also recognized that neither government nor American industry was taking sufficient notice of two emerging trends that would have long-term disruptive effects on the nation's economy: globalization and automation. At a time when many white middle-class Americans were experiencing unprecedented prosperity, she was warning her fellow citizens that their economic well-being would inevitably be overtaken by the movement of capital and jobs abroad unless they recognized the problem and did some serious sustained planning to combat it. Planning was also needed to address the challenge of automation, which was already making inroads in the American economy and which, if left unchecked, she thought would lead to chronic joblessness and poverty. Today's headlines confirm her prescience.

Failure to tackle these problems in a systematic way would have enormous consequences, politically and socially. Without a vibrant economy that worked for everyone, she feared the country's democratic institutions would be imperiled. Maintaining a standard of living that promoted "justice for all" was vital if democracy, self-government, and social cohesion were to survive.[3]

To reach these goals, she consistently argued that business, government, and labor needed to work together in a coordinated fashion to devise an economy built on the twin pillars of full production and full employment that could meet the "infinite" needs of the world.[4]

THE UNITED STATES' ECONOMY

No government or civilization could or should endure which cannot provide people with an opportunity to earn the necessities of life.

(It's Up to the Women)

Even to dream, one must have a basis of economic security, and the dream is worth little if it can not provide that. Devotion to democracy, devotion to liberty, what we call patriotism, depends upon the realization of such conditions in our country as really give us the opportunity and hope for future dreams.

(My Day, January 6, 1941)

The result of a laissez-faire policy seems to be so harmful to our whole civilization, and will hurt particularly the weak and blameless.

(My Day, August 22, 1946)

We have allowed certain practices which are wrong to grow up among the business men of the country. Had we always insisted that the men take into account the human element that entered into business there would not be the fight there is to-day over what is the fair standard of living for every one in this country. Had we said we will have none of your products which are produced without consideration of these standards which we believe in, somehow or other by now it would all have been changed.

(*It's Up to the Women*)

We are usually averse to government planning. We consider it a menace to our free-enterprise system. And yet we have no respect for an individual who does not foresee the eventualities of non-planning in his own life or business.

(My Day, March 28, 1946)

When certain things become unprofitable to do at home, instead of meeting the problem by putting up a high tariff wall, we will have to meet the challenge by changing our own economy, retraining our people, bringing in new industries. All of this takes real planning.

(My Day, June 11, 1958)

If we economize too far and on the wrong things, we are apt to find that we have cut down the people's earning power to such an extent that we have injured our whole economy rather than benefitted it.

(My Day, March 6, 1946)

There is absolutely no use in producing anything if you gradually reduce the number of people able to buy even the cheapest products. The only way to preserve our markets is to pay an adequate wage.

("The State's Responsibility for Fair Working Conditions," *Scribner's Magazine*, March 1933)

No matter what we can afford to buy, we cannot afford to buy at the expense of the health and strength of our fellow human beings.

(*It's Up to the Women*)

Much of what was done in the Depression, no matter how wasteful you think it was, left us a heritage of forests, soil conservation, roads, buildings and dams.

(My Day, April 20, 1957)

The free enterprise people who cry out loudly are the ones who want to grab freely and who will not acknowledge that in order to make things really free for the majority of the people there probably must be some control vested in government which will prevent the strong from removing all opportunities from the weak.

(*If You Ask Me*, 1946)

It has been a long fight to put the control of our economic system in the hands of the government, where it can be administered in the interests of the people as a whole. Now Congress, under the influence of powerful lobbies, is rapidly trying to return control to big business. It may be that individual Congressmen do not realize just what they are doing, but they are heading us straight for inflation and accepting the old "boom and bust" ideas, instead of sticking to the plan of ironing out the peaks and the valleys and trying to keep us on a fairly even keel.

(My Day, April 30, 1946)

Sometimes it looks as if those who depend the most upon the Federal government are not the ordinary citizens but the great corporations.

(My Day, March 19, 1957)

For many people a revival of business is not regarded primarily from the point of view of the re-employment of men, but is thought of as an opportunity to make more money. Re-employment is spoken of as a thing which will naturally follow business revival.

I confess to some trepidation where re-employment is not the first consideration; where it is not fully realized that earned wages for more people is our best way to create better markets, and thus to establish our economic system on firm basis. If we put the making of money first . . . we shall have the cart before the horse.

("Intolerance," *Cosmopolitan*,
February 1940)

Our old economic theories are not going to prove a successful guide to the future. We are becoming dimly aware of that even now. We must think and plan on a broader scale than ever before, on a scale that goes beyond our own borders, a scale that encompasses the world.

(*Tomorrow Is Now*)

If we want to keep capitalism, we in this country have got to learn that there must be real sharing, a real understanding, between management and labor. They must plan together because their interests are really identical.

(Address to Americans for Democratic Action,
April 1, 1950)

We must maintain a standard of living which makes it possible for the people really to want justice for all, rather than to harbor a secret hope for privileges because they cannot hope for justice.

(*The Moral Basis of Democracy*)

Real prosperity can only come when everybody prospers.

(My Day, March 19, 1936)

TECHNOLOGY

Once, however, the benefits of an economic revolution can be understood, once the first signs of improvement are discerned, there can be no turning back. After man discovered the benefit of a house he did not revert to a cave.

(*Tomorrow Is Now*)

It is a new industrial revolution that we are pioneering. The eyes of the world are on us. If we do it badly we will be criticized and our way of life downgraded. If we do it well we can become a beacon light for the future of the world.

(*The Autobiography of Eleanor Roosevelt*)

Either science will control us or we will control it. That is the sum and substance of the matter. By becoming its master we can build the kind of world we want to have.

(*Tomorrow Is Now*)

Automation, which could be a blessing, looms as a new fear because it is nobody's business to do the kind of planning that would eliminate fear and allow us to profit by new methods.

(*Tomorrow Is Now*)

It is true that machines have taken over the work of human hands to a great extent, but the real problem before us is how to make the work of the machines a benefit to human beings and not a detriment.

(My Day, August 7, 1941)

With increasing automation there are fewer and fewer jobs of the familiar kind for our labor force. . . . It is happening not only to the man who works with his hands but to the white-collar worker. In office after office, throughout this country, employee after employee is being eliminated and his desk remains empty. And there is no place else for him to go.

(*Tomorrow Is Now*)

UNEMPLOYMENT

We seem to forget that the unemployed are individuals, human beings with all the tastes and likes and dislikes and passions we have ourselves. When we meet them as individuals our feeling is entirely different, but as a group we talk about them as though they were so many robots.

(My Day, April 29, 1936)

The heaviest burden any people can carry is unemployment.

(*Tomorrow Is Now*)

Our recurring and increasing unemployment problem, which lies at the very heart of our economy, can be solved only by a recognition of new conditions, by a wide program for the teaching of new skills.

The needs of labor and the needs of the world will not be met by the panacea of shorter hours and higher wages. They will be met by modernizing the skills of our labor force so that they will be equipped for the new kinds of work required in a new world.

(*Tomorrow Is Now*)

If you have no unemployment you have no drag on your economy.

(*Tomorrow Is Now*)

We don't need to eliminate workers, we need to create jobs.

("Women in Politics," *Good Housekeeping*, April 1940)

POVERTY

Our people are resourceful and self-reliant, but those qualities do not always bring them the plenty which seems to be so close at hand.

(*My Day*, November 26, 1937)

Poverty is like a giant infection which contaminates everything—we know that unless we can eradicate it by the use of all our new scientific and economic materials, it can in time destroy us.

(*Tomorrow Is Now*)

How can there be any lasting faith between individuals when members of one group are constantly conscious of the fact that their life is so conditioned that they can never hope to leave misery behind them, and attain enduring independence sufficient to provide them with a simple and decent method of livelihood?

(*It's Up to the Women*)

Over and over again, it has been proved that better conditions in the lower-income brackets mean more profit at the top, and yet it is hard to convince many people in our country that this is actually so.

(My Day, August 3, 1946)

No government has a right to let its people starve.

(*Tomorrow Is Now*)

WORK

Every human being must have an equal chance to earn a living according to the opportunities open in his area of the world.

(My Day, July 24, 1945)

The ability to work and feel useful is the only thing that will make life worthwhile.

(My Day, April 2, 1946)

That a worker should earn a living wage and that it should be adequate to permit some recreation, as well as the bare necessities of life, is one of our highest aspirations.

(My Day, July 5, 1954)

The important thing in choosing an occupation is to choose something which you enjoy doing so that your reward may not be only in your salary, but may lie also in joy in your work.

(It's Up to the Women)

All work, of whatever nature, should be done to the limit of the abilities of the man who is performing the job. I do not believe that any man who is not working to the full extent of his ability can really get satisfaction out of his work.

(My Day, May 6, 1946)

Each one of us, in our own way, must be a productive member of society. If our job is to scrub the floor and we do it with the spirit of the artist or the skilled worker, in our own little sphere we have done our part of the world's production.

(My Day, March 22, 1946)

It is a most difficult thing to do a particular job well and still keep enough perspective to be able to see it in relation to the work other people are doing.

(My Day, September 4, 1936)

The man or woman who can be the center of an organization, who can keep his finger on all the lines that flow out, without interference or too close a follow-up, and who in addition can remain himself an oasis of calm with time to listen and think, will be a pearl above price in any organization.

(My Day, September 4, 1936)

It is always better for people working in an office or in the home to work with you and not for you.

(It's Up to the Women)

There is always room at the top but to arrive there and to stay there requires exceptional industry, exceptional qualifications and the courage to think along original lines.

(My Day, September 4, 1936)

It has been my observation through many long years that frequently the man who thinks he is throwing away his career because he believes in something and acts on his belief, in the end makes his career.

(My Day, December 7, 1936)

Where the scientists are concerned they are a little appalled at where their science has led them. They can destroy civilization and they wish they couldn't. But their whole scientific thought obliges them to go on studying, testing, learning because that is the way to advance.

(My Day, February 7, 1950)

All creative people—scientists, artists or educators—do their best work when they can explore all avenues of knowledge without fear.

(My Day, April 8, 1950)

For production to have real value, it must also come up to certain standards. Just to multiply the things in the world, unless they meet the needs for which they are produced, would be a rather stupid procedure.

(My Day, March 22, 1946)

It is not more vacation we need—it is more vocation.

(*Tomorrow Is Now*)

I do not suppose, however, that any really good work is ever lost. Somewhere the seed remains and the influence is felt in the future.

(My Day, March 14, 1941)

LABOR AND LABOR UNIONS

The workers of the country represent the great mass of the producing and consuming public.

(My Day, March 7, 1951)

A dangerous thing, it seems to me, when any one part of this country suggests that it may be a better place for this or that merely because it has cheaper labor to offer. That never has been a thing, as far as I know, that we have been proud of. When we could not get enough labor we threw our doors open to labor from other countries to open up our mines, to build our railroads, and start our industries. Even then we did not boast that this was cheap labor. It was labor that we needed and we got it as it came.

(My Day, September 23, 1936)

The only thing that can be done to help any workers to obtain proper working conditions and to get better wages is to organize.

(If You Ask Me, *Ladies' Home Journal,* June 1944)

I am opposed to "right to work" legislation because it does nothing for working people, but instead gives employers the right to exploit labor. Not only does it do nothing for working people, but it robs them of the gains they have made over more than a half century of bitter struggle for betterment.

("Why I Am Opposed to 'Right to Work' Laws,"
AFL-CIO American Federationist, February 1959)

The point made by the people who advocate "right to work laws" is that no one should be forced to join a union. That is certainly incontrovertible. But unions have been established as a protection to labor, and without the protection of the union, even non-unionized labor probably would have a much harder time, since the gains made for labor all have been achieved through the efforts of unions. No one man alone has any strength, but when he is joined in a union, he is able to make advances and obtain things for his membership. Once something is obtained for union members, even the non-union workers who may be employed in the plant have to be given the same advantages.

Therefore, the man who does not join a union is benefitting from the support of the union without taking on the responsibility of giving the union his support. This is not a good position for any worker to take. . . .

(My Day, September 22, 1956)

The people will be angry with the industrial leaders who have been lacking in vision for so long, and they will feel, as I do, that these industrialists have shown themselves incapable of real leadership in the economic and moral field in this great world crisis. Nevertheless, the industrialists are not going to be blamed alone. When the man in the street is really uncomfortable, he is going to blame also the leadership of any labor group which brings about his discomfort.

(My Day, May 9, 1946)

A strike will bring serious resentment and the same lack of consideration of the rights and wrongs at issue which always results when the public is seriously inconvenienced.

(My Day, March 6, 1946)

Desperate men don't strike. . . . During the depression there were few strikes. A strike is a sign of a worker's faith that he can better his condition.

(*You Learn by Living*)

MONEY

The greatest benefit which money can bring to any of us is freedom of mind, so that we may think of things which will lighten the burdens of others and a freedom of time in which to do some of these things.

(*It's Up to the Women*)

One is just a little suspicious when people say that they want no profit, yet obviously have something which might be produced for great profit!

(My Day, August 1, 1946)

I wonder whether our greed makes it impossible for us to profit by the lessons of the past.

(My Day, March 22, 1946)

It is pointless to discuss what is "enough" money. No two people would be likely to agree on what is "enough."

(*You Learn by Living*)

INTERNATIONAL TRADE

It is enlightened selfishness to build up the ability of other nations to a higher standard of living. We thus produce wider markets for ourselves as well as the rest of the world.

(My Day, July 17, 1942)

The sooner labor costs can be brought into closer rela-tionship in the various countries of the world, the better it will be for our own trade situation.

(My Day, July 15, 1957)

We cannot be an island of prosperity in the midst of a world of misery.

(My Day, February 21, 1946)

Five

WOMEN, MEN, AND GENDER

WOMEN WERE ONE OF ELEANOR ROOSEVELT'S core constituencies. Their problems, needs, and status were central to her activism, and during her lifetime she was recognized as one of their principal champions. Her success in that role stemmed in part from her ability to blend the traditional values of marriage and family with the then-controversial notion that women were entitled to lives and careers of their own. She herself embodied that duality, building a highly visible public career while maintaining a difficult marriage.

Having overcome significant professional and personal obstacles, she concentrated on expanding opportunities for women. For example, during the Great Depression she worked to include unemployed women in federal jobs programs and supported the appointment of well-qualified women to important positions in the federal government. To encourage newspapers to employ women, she stipulated that only female journalists could attend her weekly White House press conferences. When women flooded into the workforce during World War II, she advocated for equal pay for equal work, as well as for day care centers and community kitchens to ease the burden on female war workers with families. In the late 1940s, as women workers were displaced to make room for returning veterans, she defended their right to continued employment. She also continued to publicize the achievements of successful women, believing that one woman's success meant greater opportunities for all females. A pathbreaker herself, she felt that responsibility keenly, worried that if she failed at any professional task, women coming after her would have fewer options. She called on all women, regardless of whether they worked outside the home, to be loyal and generous rather than disparaging or judgmental when a female's accomplishments were acknowledged or celebrated.

Although she was a consistent and forceful advocate for women's rights, ironically Eleanor never completely supported the Equal Rights Amendment. Initially she feared it would weaken earlier protective legislative provisions for women in industry, such as the maximum hours of work and task exemptions, that had taken years to achieve. Instead she favored eliminating prohibitions against women's rights and activities in existing state laws. However, she did rethink her position in the 1940s and 1950s as women made inroads in male-dominated trade unions. Her postwar work with the UN Commission on Human Rights and the UN

Commission on the Status of Women also influenced her, but while her thinking evolved, she never wholeheartedly favored the Equal Rights Amendment.[1]

However, she consistently supported the right of women to make their own choices in life, because she recognized that they, like men, were individuals, "influenced by their environment and their experience." As for women's relationships with men, she believed the sexes complemented each other and neither should dominate at home or in the workplace.[2]

Because she recognized and appreciated women's individuality, she knew the chance of getting all of them to agree on any issue or course of action was remote. However, she left open the possibility that if they did, "something historically important will happen."[3]

GENDER

It is the person and not the sex which counts.
(If You Ask Me, *Ladies' Home Journal,* January 1942)

I have never wanted to be a man. I have often wanted to be more effective as a woman, but I have never felt that trousers would do the trick!
(If You Ask Me, *Ladies' Home Journal,* October 1941)

I could never see the use in a woman thinking or acting like a man. Men and women have always complemented each other; they do not need to imitate each other.

(My Day, December 28, 1951)

As a rule women know not only what men know but much that men will never know.

(My Day, March 6, 1937)

WOMEN

We women should be loyal and generous to each other. We cannot all do the same things, but we can admire other women when they do good work in the occupations which they feel they can undertake.

(My Day, October 22, 1943)

Women must become more conscious of themselves as women and of their ability to function as a group.

("Women in Politics,"
Good Housekeeping, April 1940)

It is perfectly obvious that women are not all alike. They do not think alike, nor do they feel alike on many subjects. Therefore, you can no more unite all women on a great variety of subjects than you can unite all men.

("Women in Politics,"
Good Housekeeping, April 1940)

It will always take all kinds of women to make up a world, and only now and then will they unite their interests. When they do, I think it is safe to say that something historically important will happen.

("Women in Politics,"
Good Housekeeping, April 1940)

There is no doubt that we women must lead the way in setting new standards of what is really valuable in life.

("What I Hope to Leave Behind,"
Pictorial Review, April 1933)

I do not really know what are the best years of a woman's life, because it depends so much on how she develops. If she is able to learn from life to get the best out of it at all times, then probably at whatever age she is those years will be the best she has had.

(If You Ask Me, *McCall's,* March 1951)

WOMEN AND POLITICS

Women who want to lead must stand up and be shot at. More and more they are going to do it, and more and more they should do it.

(*Australian Women's Weekly,*
September 11, 1943)

[Women] must stand or fall on their own ability, on their own character as persons.

(*The Simmons Program*, September 18, 1934)

There is a great consciousness of feminism only when there are many wrongs to be righted.

(White House press conference, December 17, 1944)

When anyone, man or woman, goes into politics, I believe one has to develop a pretty tough skin and take for granted one will be treated no more gently than any other candidate.

(If You Ask Me, *McCall's*, March 1962)

Paradoxically the world expects a better job from a woman in office than from a man and although it seems unfair, and is unfair, I am glad that it is so. In a manner of speaking it is a safeguard for women. It will put in office and keep in office only those who are preeminently fitted for the job.

("A Place for Women in Politics," *Women's Home Companion*, June 1932)

There is no doubt in my mind that women work differently than men, and they will put a different emphasis on certain issues. . . .

If women really understand the issues they will probably talk more effectively to their neighbors than any of the men, especially if the issues are such that they affect their daily lives.

(My Day, December 7, 1955)

I think it is fairly obvious that women have voted on most questions as individuals and not as a group, in much the same way that men do, and that they are influenced by their environment and their experience and background, just as men are. . . .

("Women in Politics,"
Good Housekeeping, March 1940)

There are liberals and conservatives among the women as well as among the men.

("Women in Politics,"
Good Housekeeping, March 1940)

Women will belong to political parties; they will work in them and leave them in much the way that men have done. It will take some great cause that touches their particular interests to unite them as women politically, and they will not remain united once their cause is either won or lost.

("Women in Politics,"
Good Housekeeping, April 1940)

There is a tendency for women not to support other women when they are either elected or appointed to office. There is no reason, of course, why we should expect

any woman to have the support of all women just because of her sex; but neither should women be prejudiced against women as such. We must learn to judge
other women's work just as we should judge men's
work, to evaluate it and to be sure that we understand
and know the facts before we pass judgment.

("Women in Politics,"
Good Housekeeping, April 1940)

WOMEN'S RIGHTS

The battle for the individual rights of women is one of
long standing and none of us should countenance anything which undermines it.

(My Day, August 7, 1941)

I know of no better way, however, to educate women to
their responsibilities as citizens than to give them civil
and political rights on an equal basis with men.

(My Day, November 18, 1946)

The advance of women is a fairly good measure of the
advance of democracy. Their advance is slow just as
is the growth of democracy. Each step taken forward,
however, is a real step toward freedom and justice.

(My Day, March 15, 1948)

Every now and then I am reminded that even though the need for being a feminist is gradually disappearing in this country, we haven't quite reached the millennium.

(My Day, February 23, 1945)

There are many ways in which women can be discriminated against which are not going to be remedied by legal steps alone, and that also holds good in discrimination for other reasons. The hearts of men will have to change, not just the laws under which we live.

(My Day, August 10, 1957)

WOMEN AND WORK

If holding a job will make a woman more of a person . . . then holding a job is obviously the thing for her to do.

("What I Hope to Leave Behind,"
Pictorial Review, April 1933)

There is no question about the validity of the doctrine of equal pay for equal work. It is surprising to me that we still have to fight for such an obvious right. A woman who does the same work as a man should without any question receive equal pay. This is justice and should not be overlooked in any country which claims to treat its people without discrimination.

(Undated draft statement circa 1951
for *Red Tape: The Civil Service Magazine*)

If women do the same work I have always believed that they should receive the same pay.

(If You Ask Me, *Ladies' Home Journal*, March 1944)

I think it does need constant watchfulness on the part of women's groups and of individual women to protect their economic gains.

(*If You Ask Me*, 1946)

A woman will always have to be better than a man in any job she undertakes.

(My Day, November 29, 1945)

MEN AND WOMEN

I'm quite willing to grant that men, on the whole, have accomplished more in the art of oratory than have women, but if you want to get work done quickly, oratory is not half so important as putting your thoughts clearly, taking up as little time as possible, and never speaking unless you have something that really needs to be said.

(My Day, February 13, 1946)

Men and women can be equal but they cannot be identical. They always will have different functions, and even though they do the same things they often will do them differently.

(My Day, May 25, 1951)

We cannot change the fact that women are different from men. It's true that some women can do more than men, and some can do men's jobs better than men can do them. But the fact that they are different cannot be changed, and it is fortunate for us that this is the case. The best results are always obtained when men and women work together, with the recognition that their abilities and contributions may differ but that, in every field, they supplement each other.

(My Day, June 1, 1946)

[Men and women] are both needed in the world of business and politics to bring their different points of view and different methods of doing things to the service of civilization as individuals, with no consideration of sex involved; but in the home and in social life they must emphasize the difference between the sexes because it adds to the flavor of life together.

("Women in Politics,"
Good Housekeeping, April 1940)

Women have one advantage over men. Throughout history they have been forced to make adjustments. . . . [T]he result is that, in most cases, it is less difficult

for a woman to adjust to new situations than it is for
a man.

(*You Learn by Living*)

MARRIAGE

In many respects marriage will always remain a gamble. If we could take the gamble out of it, it would be far less interesting.

("Ten Rules for Success in Marriage,"
Pictorial Review, December 1931)

A marriage is a partnership in which success and happiness are achieved by joint decisions and joint actions.

(*Book of Common Sense Etiquette*)

No two people really know each other until they have been married for some time.

("Should Wives Work?,"
Good Housekeeping, December 1937)

No one can decide for you what will be a good marriage and under what circumstances you would be willing to join your life with that of another human being.

(*My Day*, July 16, 1952)

I think we ought to impress on both our girls and boys that successful marriage requires just as much work, just as much intelligence and just as much unselfish devotion, as they give to any position they undertake to fill on a paid basis.

(My Day, March 29, 1941)

Real loving means work, thinking of each other day in and day out, unselfishness, and effort to understand the growth of the soul and mind of the other individual, and to adjust and complement that other person day by day.

(My Day, October 20, 1939)

I think anything connected with the home is as much the husband's work as the wife's. This silly idea that there is a division in housework seems to me foolish, when very often the wife earns money outside the home as well as the husband. . . . The kind of man who thinks that helping with the dishes is beneath him will also think that helping with the baby is beneath him, and then he certainly is not going to be a very successful father.

(If You Ask Me, *Ladies' Home Journal,* January 1945)

The successful wife is many women; lover, housekeeper, mother (not only to her children, but in many ways to her husband as well), a listening ear, a sympathetic voice, a tender and comforting hand, a first-aid nurse, a stimulant to cheerfulness on gloomy days, a gracious hostess, and frequently one who shares with her husband the task of earning the family livelihood. I often have thought that less is expected of the president of a great corporation than of an American wife.

(*Book of Common Sense Etiquette*)

I think people are happier in marriage when neither one is the boss, but when both of them are willing to give as well as take.

(If You Ask Me, *Ladies' Home Journal,* September 1944)

All human beings have failings, all human beings have needs and temptations and stresses. Men and women who live together through long years get to know one another's failings; but they also come to know what is worthy of respect and admiration in those they live with and in themselves.

(*The Autobiography of Eleanor Roosevelt*)

DIVORCE

A marriage without love is intolerable, and a home without love is a poor place for a child to grow up.

(If You Ask Me, *Ladies' Home Journal,* February 1942)

Divorce is at best an unhappy affair, the admission of failure and the termination of a program which began with high hopes and expectations for a lifetime.

(*Book of Common Sense Etiquette*)

Today many seem to think that marriage is like a position in employment, which one can leave when everything does not go well. We should think of it as a permanent, lifetime job.

("Ten Rules for Success in Marriage," *Pictorial Review*, December 1931)

MOTHERHOOD

Women as mothers are constantly projecting themselves into the future and are therefore deeply concerned with every phase of the preservation and development of life.

(My Day, February 22, 1946)

As citizens, if mothers would just get together and agree on what they wanted, I think they would find that their influence and power was very much greater than they had ever dreamed. A mother has a twofold responsibility: that of using her own influence as much as her duties will permit to bring about the end which she desires, and the power of her example on the future citizens.

(*The Pond's Program*, February 3, 1933)

It is the mother's responsibility to see that questions which affect the public good in the community and in the state and in the country are discussed at the table and around the fire so that her children will really become interested in talking about something more than the small gossip about their friends and their own plans and immediate concerns.

(*The Pond's Program*, February 3, 1933)

BIRTH CONTROL

I believe people should be given freedom to make their own choices. Those who do not want to hear or know certain things should not be forced to do so. Neither should it be made impossible for those who desire certain kinds of information to obtain it.

(My Day, February 21, 1950)

I am for birth control, particularly in places having severe problems of overpopulation and insufficient food. I can quite understand, however, that there are reasons why certain people do not believe in birth control, and I feel that people should be allowed their own decisions in this matter.

(If You Ask Me, *McCall's*, September 1958)

I think that very few marriages are entirely happy if people voluntarily give up having children. Children are both a joy and a responsibility. They bring both pleasure and sorrow into the lives of their parents, but they are a tangible expression of the real love that exists between man and wife; and if you could have children and deliberately avoided it, I think something would be missing in the marriage relationship between young people.

(If You Ask Me, *Ladies' Home Journal,* August 1945)

CHILD CARE

I think it is absolutely vital for the good of future generations that nursery schools be made available to care for the children of working mothers during the entire day. The next generation will seriously suffer from both the physical and mental standpoints if this is not done.

(If You Ask Me, *Ladies' Home Journal,* April 1946)

Instead of having government-run nurseries it is better to have cooperative nurseries where parents arrange to pay some small amount and have a say in how the nursery is run. If these nurseries need a subsidy, it should come from local organizations or government.

(If You Ask Me, *McCall's,* June 1955)

Six

YOUTH AND EDUCATION

ELEANOR ROOSEVELT IDENTIFIED WITH CHILDREN AND young people. The product of a rigid and difficult childhood (orphaned by the age of ten, she had been raised by her grandmother), she understood and sympathized with the problems of the young, and they in turn felt they had an ally in her. Eleanor offered more than a sympathetic ear or a shoulder to cry on, though. She actively worked to meet their needs, because she believed that their futures and that of democracy were inextricably intertwined.

Her involvement with children began before her marriage, when she taught dancing and calisthenics at a New York City settlement house. Before and after she became First Lady, she supported legislation to abolish child labor and provide better health care and nutrition for children and mothers. Concerned that the ravages of the Great Depression would cause young people to abandon democracy, she was instrumental in the founding of the National Youth Administration (NYA), a federal program that helped young people stay in school and provided job training for those who had graduated. During World War II she advocated for the needs of the young people serving in the military, with special emphasis on the discriminatory conditions African Americans endured. As the war wound down, she lobbied FDR for legislation to guarantee returning veterans jobs and education. In the postwar era she made a point of speaking to as many youth groups as possible to encourage and challenge them to fulfill the promises of democracy.

Eleanor's activism on behalf of the young also included strong support for education in general and teachers in particular. She considered teachers "among the most important people" in the nation because of the role they played in teaching and modeling democratic values for young people.[1] Those values included two skills she considered vital for individuals living in a rapidly changing world: how to acquire knowledge and how to think for themselves. Children given that foundation, she believed, would grow up to be individuals who would exercise the responsibilities of citizenship intelligently.[2]

CHILD-REARING

All children, it seems to me, have a right to food, shelter and equal opportunity for education and an equal chance to come into the world healthy and get the care they need through their early years to keep them well and happy.

("Insuring Democracy,"
Collier's, June 15, 1940)

I have always found that children were happier if they were allowed to have something which made them distinctly like all other children, than if they were obliged to express the individuality of their elders.

(*It's Up to the Women*)

It does us elders much good to see ourselves through the eyes of a child.

(My Day, January 17, 1946)

Perhaps what all of us need in our relationships with children is a clearer memory of our own childhoods.

(If You Ask Me,
McCall's, August 1962)

I think it is essential that you should teach your child that he has an intellectual and spiritual obligation to decide for himself what he thinks and not to allow himself to accept what comes from others without putting it through his own reasoning process.

(*You Learn by Living*)

A child who has been treated with real respect, who has a feeling that his elders expect certain standards even from a young member of the family, will behave with astonishing maturity. A child who feels the basic acceptance that goes with respect, and knows he is trusted because he is accepted, will achieve remarkable ability in acquiring self-control and consideration for others.

(*You Learn by Living*)

A child responds naturally to high ideals, and we are all of us creatures of habit.

("Good Citizenship: The Purpose of Education," *Pictorial Review*, April 1930)

It is when you set standards for youngsters and do not live up to them yourself that youngsters begin to question the validity of the standards.

(My Day, July 27, 1956)

If you live what you believe, your children will believe it, too.

(*You Learn by Living*)

The child who is aware that his parents do not tell him the truth will assume that the practical method is to lie.

(*You Learn by Living*)

The great thing, I think, is to make children feel that they are needed and actually belong in the family.

(If You Ask Me, *McCall's*, October 1952)

Children get out on their own sooner now and they have to be trusted to decide things for themselves. That is why I think mothers should explain the reasons behind their direction, whenever possible. At the same time, a child should know that if it is told to do something, it must do so, even if the reasons can't be explained until later on.

(*The Pond's Program*, May 5, 1937)

I believe very strongly that it is better to allow children too much freedom than too little; it is better for them to get their feet wet than to be told at the age of fifteen to put on their rubbers. They should be old enough by that time to take care of themselves and if they prefer to get their feet wet, they should be allowed to do so.

(*It's Up to the Women*)

There is no question in my mind that responsibility is a good thing, but it can not be administered artificially, so we can only be grateful when our children have it thrust upon them naturally.

(My Day, January 6, 1936)

It sometimes is hard to allow young people to learn from their own experience, but in the long run, I think, it is probably the only way they ever do learn.

(My Day, August 15, 1957)

Children rush from school to basketball games or television programs. At the age of the keenest curiosity and the widest interest, they are, it seems to me, missing the boat altogether. Anyone who has watched small children at play will observe that, for the most part, the play is earnest. They are happiest when it is creative. They want to accomplish something, to make something. But when they are permitted, even encouraged, to accomplish as little as possible, this early innate impulse begins to fade.

(*Tomorrow Is Now*)

A special contribution to the life of children is to have time to be leisurely with them.

(*My Day, September 3, 1954*)

People should grow naturally with their children but be ready to let them go when the time comes.

(*You Learn by Living*)

All parents, I think, feel that they haven't always been wise and that they are in some way responsible if their children suffer later on because of traits of character which might have been obliterated when they were young. The only way I think that parents can meet that is to accept the fact that no human being is all-wise; no human being always lives up to the best that he is capable of, all the time. Failures come to all people.

(*If You Ask Me,
Ladies' Home Journal, March 1947*)

TEENAGERS

Our great trouble is that we tend to bring up our children for the world in which we ourselves have lived. We rarely take into consideration the changes which have come about in our own lifetime and which are likely to come about before our children are fully grown. Perhaps we handicap them unnecessarily by too much anchoring to the old and too little preparation for the new.

(My Day, July 7, 1942)

We must all realize, I think, that between generations there is a tremendous gulf and that each new generation sets up its own standards as the result of contact with its own contemporaries.

(*It's Up to the Women*)

Young people need help, but they do not need criticism or interference.

(*It's Up to the Women*)

After the early years when a parent encompasses the whole of a child's world, the youngsters are apt to be greatly influenced by what their contemporaries and outside contacts are saying. The atmosphere and the example that they have at home, however, is, I think, always the greatest influence in forming their characters and permanent mode of thought.

(My Day, September 5, 1936)

We cannot expect to shape [young people's] ideas or in fact to shape the world which they are going to live in. All we can hope to do is to make them feel that they must be honest with themselves and honest with those they care about, living up to what they think is right and remembering always that we are not the judges of what is going to make the world a better place for them to live in. That we must leave to the judgment of the generation that comes after us.

(*It's Up to the Women*)

There is something about a big group of youngsters facing one at a school assembly which always makes me want to be working with them again—I suppose they appeal to one's imagination because they have so many possibilities before them.

(My Day, October 19, 1937)

As I looked at the eager sea of young faces before me, I had the curious feeling that youthful audiences so often give me. There is a desire to know and to hear and yet a veiled challenge, an apparent question in their minds as to whether you have anything to give them.

(My Day, March 27, 1936)

I always approach the younger generation with the hope that here I am going to find another friend and new interests and new insights.

(My Day, January 9, 1957)

Snap judgments which one must give on a hundred and one different questions from young people, are one of the things I think most difficult, for there is a great responsibility to be honest with youth. . . .

(My Day, November 18, 1937)

Perhaps the older generation is often to blame with its cautious warning: "Take a job that will give you security, not adventure." But I say to the young: "Do not stop thinking of life as an adventure. You have no security unless you can live bravely, excitingly, imaginatively. . . ."

(*The Autobiography of Eleanor Roosevelt*)

Sometimes I think that the younger generation has seemed to make more mistakes than the older ones for the simple reason that they have not been given such black and white values.

(My Day, March 29, 1958)

I am convinced that one of the reasons why our young people today feel uncertain is that they are not being trained to examine questions, to decide for themselves on a method of action—*and then to act*. The result is that they may too easily be acted upon.

(*Tomorrow Is Now*)

I am often told by young people in the universities, and told rather resentfully, that it's unfair they have to cope with a world they never made. But this is the common lot of man. What I think they tend to forget is that the world they hand on to their children will be a world they *have* made. If they do not like their inheritance, at least they have every opportunity to improve on it. And that, I feel most strongly, is the great adventure of life, an adventure renewed for every generation.

(*Tomorrow Is Now*)

That is what encourages me about the younger generation. They take an interest in matters beyond their own immediate surroundings because they have seen what an impact the various parts of the world have on one another. We must hope that they can pass on their experience and interest to the next generation, which we hope shall not have to fight a war to obtain it.

(My Day, January 25, 1950)

In the hands of the young people lies the future of this country, perhaps the future of the world and our civilization. They need what help they can get from the older generation and yet it must be sympathetically given with a knowledge that in the last analysis the young people themselves must make their own decisions.

("I Want You to Write to Me,"
Women's Home Companion, August 1933)

When you are young it is nice I think to feel that somewhere in the background there are some older people

to whom you can turn. They are a kind of bulwark between you and the future.

(My Day, August 9, 1937)

Only as our youth dreams in terms of living will this
world see much change in philosophy or action.

(My Day, October 26, 1936)

Youth and hope will win out I am sure.

(My Day, February 18, 1936)

EDUCATION

We need to change our attitude in this country toward
learning and knowledge, its value, and the respect due
to those who take the trouble to learn.

(My Day, March 3, 1958)

The average home in this country will buy a TV set before it will see to it that the children have the necessary
books to read at home which will enrich their school
courses.

(My Day, April 17, 1961)

If we look at education not as something to end at any
special point, but as a preparation whereby we can

attain the means to acquire whatever we need in life, we shall have a truer perspective on what we want to achieve through our educational system.

(My Day, December 30, 1942)

It is never enough, it seems to me, to teach a child mere information. In the first place, we have to face the fact that no one can acquire all there is to learn about any subject. What is essential is to train the mind so that it is capable of finding facts as it needs them, train it to learn how to learn.

(*You Learn by Living*)

As the volume of knowledge about our world increases, programs of study are becoming more and more complicated, and we must strive harder to give our children the kind of education that would fit them for a life that must be lived in a changing world.

(My Day, April 11, 1962)

We are not doing such a remarkably good job in education as to assume that we have done all that can be done for our young people.

(My Day, July 8, 1957)

Have we been honest and brave enough to make clear to our young people that we could not, if we would, provide the specific education that would equip them for the coming world, whose face no one knows? Have we told them bluntly that the best we can do is to give them skills, to train the mind so that it becomes a flexible tool . . . ?

(*Tomorrow Is Now*)

I have always believed that beautiful surroundings were a help in education.

(*My Day*, October 12, 1937)

There is a wonderful word, *why?*, that children use. All children. When they stop using it, the reason, too often, is that no one bothered to answer them. . . .

Every child's *why* should be answered with care— and with respect. . . . If the child's curiosity is not fed, if his questions are not answered, he will stop asking questions. And then, by the time he is in his middle twenties, he will stop wondering about all the mysteries of the world. His curiosity will be dead.

(*You Learn by Living*)

What I have learned from my own experience is that the most important ingredients in a child's education are curiosity, interest, imagination, and a sense of the adventure of life.

(*You Learn by Living*)

I have come to believe that one of the essentials of education is developing a sense of self-confidence.

(*The Pond's Program*, May 5, 1937)

Every child should be given some business education besides the ordinary standards of honesty. They
should be taught that you do not spend money until
you have it, that money represents somebody's work
or the production of some material things for which
some human beings in some way have worked. Money
is only a token but it is a token which represents real
things.

(*It's Up to the Women*)

What a child understands and retains is a lesson in
the way most of us handle the subject we are talking
about.

(My Day, January 10, 1950)

EDUCATION FOR CITIZENSHIP

We are concerned about the children before they are
born, but we should follow them through every step of
their development until the children are firmly on their
feet and started in life as citizens in a democracy.

("Insuring Democracy,"
Collier's, June 15, 1940)

What happens to our children is the concern of the
whole nation because a democracy requires a standard

of citizenship which no other form of government finds necessary.

("Insuring Democracy,
Collier's, June 15, 1940)

It takes a good home and a good school to prepare young people for citizenship in a democracy and even then they will have to go on learning throughout life.

(*If You Ask Me*, 1946)

I am convinced that every effort must be made in childhood to teach the young to use their own minds. For one thing is sure: If they don't make up their minds, someone will do it for them.

(*You Learn by Living*)

It is the duty of parents and teachers to see that into a child's consciousness gradually grow these interests of the greater world and an understanding of good citizenship and what responsibilities toward the community and the country mean.

(*It's Up to the Women*)

Whatever really interests young people and encourages the habit of reading so that they keep up not only with the current events of their own environment, but the current events of the world at large is good training for general living in a democracy.

(*My Day*, November 1, 1937)

We want to bring up children who have an understanding of the problems of the world, who have the ability to be fair and generous in their judgments and strong enough to stand by their beliefs and to work for peace in the same way that, in the past, we have worked to build up the material success of our nation and its defense when it was needed.

(*Americans of Tomorrow*, November 11, 1934)

As we remove economic pressure by giving people a minimum of security, we must, through education, see that they acquire an even more ardent desire to serve their community. . . .

(My Day, February 9, 1943)

In a democracy such as ours, the education of all the people is a vital necessity. They cannot become articulate and express their beliefs unless they can both write and speak.

(My Day, November 3, 1945)

We cannot afford to waste brains in this country. They are becoming more important to us every day. And surely financial position should not bar young people from the education which can give them positions of leadership in our nation in the future.

(My Day, February 22, 1951)

All despotism is based on education which holds people to a definite pattern and crushes individuality.

(My Day, March 21, 1936)

One of the best ways of enslaving people is to keep them from education and thus make it impossible for them to understand what is going on in the world as a whole.

(My Day, May 11, 1943)

PUBLIC EDUCATION

Our forefathers realized the importance of education to a nation, and particularly the importance of an educated people in a democracy, but we seem to have forgotten what sacrifices they made to start a public school system in this country.

(My Day, March 20, 1957)

The school houses of a community are not only for the use of the children and the educators. They should be the centers from which radiate the ideas which motivate the community.

(My Day, January 24, 1944)

Poor schools in our communities today mean poor citizens in these communities in a few years—men and women ill-prepared to earn a living, or to participate in government.

("Insuring Democracy," *Collier's*, June 15, 1940)

EXPERIENTIAL EDUCATION

All of life is a constant education.

(My Day, December 22, 1945)

Nothing we learn in this world is ever wasted and I have come to the conclusion that practically nothing we do ever stands by itself. If it is good, it will serve some good purpose in the future. If it is evil, it may haunt us and handicap our efforts in unimagined ways.

(*The Autobiography of Eleanor Roosevelt*)

Learning and living. But they are really the same thing, aren't they? There is no experience from which you can't learn something. When you stop learning you stop living in any vital and meaningful sense. And the purpose of life, after all, is to live it, to taste experience to the utmost, to reach out eagerly and without fear for newer and richer experience.

You can do that only if you have curiosity, an unquenchable spirit of adventure. The experience can have meaning only if you understand it. You can understand it only if you have arrived at some knowledge of yourself, a knowledge based on a deliberately and usually painfully acquired self-discipline.

(*You Learn by Living*)

The best training in the world is practical experience.

(My Day, April 20, 1942)

There is no such thing, from my point of view, as overeducation, nor being above any people because of formal education that you might have been fortunate enough to acquire.

(If You Ask Me,
Ladies' Home Journal, July 1948)

There is no human being from whom we cannot learn something if we are interested enough to dig deep.

(*You Learn by Living*)

Never, perhaps, have any of us needed as much as we do today to use all the curiosity we have, needed to seek new knowledge, needed to realize that no knowledge is terminal. For almost everything in our world is new, startlingly new. None of us can afford to stop learning or to check our curiosity about new things, or to lose our humility in the face of new situations.

(*You Learn by Living*)

A person must take the trouble to learn what he needs to know if you are going to solve new problems in new ways.

(Campaign address for Adlai Stevenson,
October 11, 1956)

TEACHERS

The quality which makes men or women great teachers is the ability to inspire with curiosity the youthful mind.

(It's Up to the Women)

No one has more influence in a community than a teacher.

(It's Up to the Women)

Since we decided long ago that democracy could not exist without education, we can easily see why good teachers are essential to our development.

(My Day, January 24, 1944)

We in the United States, I believe, sometimes lay too much stress on beautiful school buildings and not enough on compensating our teachers, without whom the buildings would be useless.

(My Day, May 21, 1957)

The teachers are, of course, among the most important people in our nation. Day in and day out they are at work preparing the future citizens of the U.S. . . .

It is in the classroom that many of our children get their best lessons in democracy, and the men or women teaching our children must remember that school experience is just a preparation for the wider experience of life and citizenship in a democracy.

(My Day, January 14, 1953)

In an economy where the gauge of one's importance or of one's success is the amount of money one makes, it is quite evident that if we think teaching is an important career, it must be adequately rewarded.

(If You Ask Me,
McCall's, January 1961)

We need better trained teachers, better paid teachers, teachers who can take a more active part in community life and who are given complete freedom in the expression of their opinions.

(My Day, April 19, 1943)

Seven

FAITH AND ETHICS

FAITH WAS AN IMPORTANT COMPONENT OF ELEANOR
Roosevelt's belief system, and it undergirded much of what she thought and did. A lifelong Episcopalian from a wealthy family, her early religious training focused on Bible reading, Scripture memorization, and the importance of adhering to biblical moral imperatives. She was also encouraged to develop a sense of noblesse oblige toward those less fortunate than she. Her husband's infidelity and his subsequent bout with polio forced her to reevaluate her faith. So, too, did the relationships she formed with men and women who came from other religious traditions and walks of life. Her travels, particularly her postwar trips to the Middle East and Asia, also affected her outlook.[1]

Over time, she came to see that Christianity and

many other faiths shared certain core components: belief in a higher power, a call to live ethically and authentically, and faith that allowed believers to face the future with serenity.[2]

As a result of all these experiences, Eleanor's stance toward her own faith began to change and deepen. Instead of being a set of rules and expectations, her faith evolved into a set of beliefs and ethical constructs that animated her activism. Chief among these were her convictions that all people, regardless of race, creed, or color, were worthy of respect and care and that "real religion" manifested itself most clearly in the way people lived their daily lives.[3]

FAITH AND RELIGION

To me religion has nothing to do with any specific creed or dogma. It means that belief and that faith in the heart of a man which makes him try to live his life according to the highest standard which he is able to visualize.

("What Religion Means to Me,"
Forum, December 1932)

Religion to me is simply the conviction that all human beings must hold some belief in a Power greater than themselves, and that whatever their religious belief may be, it must move them to live better in this world and to approach whatever the future holds with serenity.

(If You Ask Me,
Ladies' Home Journal, October 1941)

The fundamental, vital thing which must be alive in each human consciousness is the religious teaching that we cannot live for ourselves alone and that as long as we are here on this earth we are all of us brothers, regardless of race, creed, or color.

("What Religion Means to Me,"
Forum, December 1932)

You may never be able to decide exactly what you believe, but you can pray to and believe in a God Whose

infinite wisdom allows freedom of thought and action,
and Who gives hope that there will be a future and that
it will be good.

(If You Ask Me, *McCall's*, April 1960)

The important thing is neither your nationality nor the
religion you professed, but how your faith translated itself in your life.

(My Day, September 23, 1943)

Real religion is displayed in the way we live in our day-
by-day activities at home, in our own communities, and
with our own families and neighbors.

(My Day, July 5, 1962)

There is a crisis in the world today and it is not just a
material crisis, though it may well seem to be. It is a
crisis that deals with the spirit and the heart and the
mind of man, and spiritual leadership is as important
as a good, hard-headed practical guidance.

(My Day, August 13, 1952)

Many of us feel that there can be no permanent settlement of the problems that face us nationally and internationally without a real spiritual awakening in the
world as a whole.

(My Day, August 8, 1946)

The churches cannot become just another pressure
group. In a country like ours, where church and state

are pretty carefully separated, great emphasis will have to be laid on fundamental principles from which action springs, rather than on the specific actions undertaken by groups of individuals.

(*My Day, August 8, 1946*)

ETHICS

The building of character and the teaching of ethics are tied up with ability to think clearly and to look honestly at situations when they arise, even when they affect oneself.

(*My Day, November 26, 1954*)

No one, it seems to me, can really see his own life clearly any more than he can see himself, as his friends or enemies can, from all sides. It is a moral as well as a physical impossibility. The most one can achieve is to try to be as honest as possible.

(*The Autobiography of Eleanor Roosevelt*)

I have often heard that in order to keep a job you will have to make compromises. But make them as little as you can, and in your private life try to live in the way you really feel you want to live. It will change even the business conditions under which you may have to struggle.

(*You Learn by Living*)

People in the higher income brackets are frequently involved in criminal procedure because somewhere along
the line they failed to recognize the ethics of a situation.

(My Day, September 12, 1962)

DEATH

To preserve the love of family and friends, and the
respect and admiration of those with whom one has
worked, is to end a life of service in this world with flags
flying and a sense of a complete life.

(My Day, January 4, 1957)

Death should be calm and serene when work is done
and well done, there is nothing to regret either for those
who go or for those who stay behind. Only an inheritance or good accomplishment to be lived up to by those
who carry a loving memory in their hearts.

(My Day, April 22, 1936)

We know that life must go on and that, while those who
leave us live on in our hearts, still the business of living
must be cheerfully accomplished.

(My Day, August 22, 1946)

I think we shall have fulfilled our mission well if when
our time comes to give up active work in the world we

can say we never saw a wrong without trying to right it; we never intentionally left unhappiness where a little effort would have turned it into happiness, and we were more critical of ourselves than we were of others.

(It's Up to the Women)

I happen to think that a belief in God is really all that is necessary for the acceptance of death, since you know that death, like life, is part of God's pattern.

(You Learn by Living)

I believe that all you go through here must have some value; therefore, there must be some *reason*. And there must be some "going on." How exactly that happens I've never been able to decide. There is a future—that I'm sure of. But *how*, that I don't know. And I came to feel that it didn't really matter very much because whatever the future held, you'd have to face it when you came to it, just as whatever life holds, you have to face it in exactly the same way. And the important thing was that you never let down doing the best that you were able to do. . . . [T]hat was what you were put here to do. . . .

(This I Believe, 202)

When you cease to make a contribution, you begin to die.

(Letter to Francis Horn, February 19, 1960)

Eight

LIFE LESSONS

MUCH OF ELEANOR ROOSEVELT'S EDUCATION CAME FROM practical experience and close observation of people, places, and situations. To her, learning and living were "the same thing." She believed that "when you stop learning you stop living in any vital and meaningful sense."[1]

Among the lessons she learned were "to be adaptable, never to make an issue of little things, to remember that the objective is important but not to force everybody else into your own pattern." Worry was "foolish" because "whatever comes we have to meet it." Criticism was inevitable, but unless it was valid it was best ignored. To her the goal of life was to become an individual with "standards," "values," and "convictions" about "what is right and wrong, what is true and false, what is important and what is trivial."[2]

While life could be difficult, Eleanor believed it was

also full of love, beauty, and adventure, provided an individual was willing to live courageously. For her, being courageous meant remaining "interested in life," reaching out boldly for new experiences, and savoring every moment. "Sorrow and suffering" were inevitable, but whatever life brought, good or bad, the important thing, she thought, was to face it, learn from it, and use it as a springboard to a fuller living experience.[3]

LIFE

The great experiences of life are the same wherever you live and whether you are rich or poor. Birth and death, courage and cowardice, kindness and cruelty, love and hate, are no respecters of persons.

("In Defense of Curiosity,"
Saturday Evening Post, August 24, 1935)

Life has gradually taught me to be adaptable, never to make an issue of little things, to remember that the objective is important but not to force everybody else into your own pattern.

(My Day, February 4, 1936)

It is the living that we do that still matters.

(My Day, July 4, 1957)

Nothing alive can stand still, it goes forward or back. Life is interesting only as long as it is a process of growth.

(*You Learn by Living*)

Life was meant to be lived, and curiosity must be kept alive. One must never, for whatever reason, turn his back on life.

(*The Autobiography of Eleanor Roosevelt*)

One should keep going to a place associated with pleasant memories, even though it may evoke regrets over certain things which never can be quite the same.

(My Day, June 5, 1957)

Some people are always complaining of the present, and then looking back upon that present with a feeling that it was rather a delightful past. They hurry toward the future in the hope that it will be more enjoyable than the present, and so life slips away and the joys of the moment are rarely savoured.

(My Day, June 16, 1947)

It's a great life if you never get tired!

(My Day, January 25, 1936)

HUMAN NATURE

Without doubt human beings are the most interesting study in the world.

(My Day, January 5, 1939)

Now the only way we can judge human nature is by human behavior, and behavior is modified and changed and developed and transformed by training and surroundings, by social customs and economic pressures.

(Tomorrow Is Now)

Human relationships, like life itself, can never remain static. They grow or they diminish. But, in either case, they change.

(You Learn by Living)

The contrariness of human nature is becoming more apparent every day. When we make mistakes ourselves we hate to have other people notice them, but when the boot is on the other foot, we leave nothing unnoticed. The youngest child acts this way, and the greatest statesmen.

(My Day, March 4, 1936)

We think our way of doing things is a good way, and we have a right to believe in it until someone else proves to us that their way is better.

(My Day, May 16, 1946)

A disagreeable fact is not acknowledged until it is necessary to do so.

(My Day, August 31, 1945)

All of us throughout the world are dealing with imperfect human beings. Our problem is not to attain perfection, but to feel that we are moving forward.

(My Day, July 1, 1946)

BECOMING AN INDIVIDUAL

We are facing a great danger today—the loss of our individuality. It is besieged on all sides by pressures to conform: to a standardized way of living, to recognized—or required—codes of behavior, to rubber-stamp thinking. But the worst threat comes from within, from a man's or woman's apathy, his willingness to surrender to pressure, to "do it the easy way," to give up the one thing that is himself, his value and his meaning as a person—his individuality.

(You Learn by Living)

It is a brave thing to have courage to be an individual; it is also, perhaps, a lonely thing. But it is better than not being an individual, which is to be nobody at all.

(You Learn by Living)

I never can understand why so many people are afraid to live their own lives as they themselves think is right.

(You Learn by Living)

Making up our minds as to what gives us the greatest amount of pleasure and then working for it, is one of the satisfactions of life. Drifting along is easy to do but if we want to see a real pattern in our lives, we must take the trouble to sit down and think out not only what we want for ourselves but what we want for our families.

(It's Up to the Women)

Obviously, it requires effort to use all your potentialities to the best of your ability, to stretch your horizon, to grasp every opportunity as it comes, but it is certainly more interesting than holding off timidly, afraid to take a chance, afraid to fail.

(You Learn by Living)

Somewhere along the line of development we discover what we really are, and then we make our real decision for which we are responsible. Make that decision primarily for yourself because you can never really live anyone else's life, not even your child's. The influence you exert is through your own life and what you become yourself.

(Letter to Trude W. Pratt, circa June 1941)

I think one of the basic things to recognize is that the only valuable development is the development of an individual. If you try to change that individual so that he loses his personality, you have done something that has destroyed the most important thing about a human being, his essential difference from anybody else. Any one of us who tries to make someone over and force him into an image of what we think he should be, rather than encourage him to develop along his own lines, is doing a dangerous thing.

(You Learn by Living)

Since everybody is an individual, nobody can be you. You are unique. No one can tell you how to use your time. It is yours. Your life is your own. You mold it.

You make it. All anyone can do is to point out ways and means which have been helpful to others. Perhaps they will serve as suggestions to stimulate your own thinking until you know what it is that will fulfill you, will help you to find out what you want to do with your life.

(You Learn by Living)

Sooner or later, you are bound to discover that you cannot please all of the people around you all of the time. Some of them will attribute to you motives you never dreamed of. Some of them will misinterpret your words and actions, making them completely alien to you. So you had better learn fairly early that you must not expect to have everyone understand what you say and what you do.

(You Learn by Living)

Remember always that you have not only the right to be an individual; you have an obligation to be one. You cannot make any useful contribution in life unless you do this.

(You Learn by Living)

VALUES

It's your life—but only if you make it so. The standards by which you live must be your own standards, your own values, your own convictions in regard to what is right and wrong, what is true and false, what is important and what is trivial. When you adopt the standards and the values of someone else or a community or a pressure group, you surrender your own integrity. You become, to the extent of your surrender, less of a human being.

(You Learn by Living)

To be mature you have to realize what you value most. It is extraordinary to discover that comparatively few people reach this level of maturity. They seem never to have paused to consider what has value *for them*. They spend great effort and sometimes make great sacrifices for values that, fundamentally, meet no real needs of their own. Perhaps they have imbibed the values of their particular profession or job, of their community or their neighbors, of their parents or family. Not to arrive at a clear understanding of one's own values is a tragic waste. You have missed the whole point of what life is for.

(You Learn by Living)

A mature person is one who does not think only in absolutes, who is able to be objective even when deeply stirred emotionally, who has learned that there is both good and bad in all people and in all things, and who walks humbly and deals charitably with the circumstances of life, knowing that in this world no one is all-knowing and therefore all of us need both love and charity.

(If You Ask Me, *McCall's*, October 1953)

In today's life *you must have convictions on basic questions.* You must make up your mind on where you stand. In the company of your own peers you should be prepared to state where you stand and defend your opinion. It is not enough to say, "I do not agree at all." You must be able to say why.

(*You Learn by Living*)

Young or old, in order to be useful we must stand for the things we feel are right, and we must work for those things wherever we find ourselves. It does very little good to believe something unless you tell your friends and associates of your beliefs.

(My Day, May 7, 1945)

When you understand yourself clearly it is easier for you to understand more clearly the people whom you love.

(*You Learn by Living*)

ADVENTURE

You cannot meet a challenge till you know what the challenge is.

(My Day, November 23, 1960)

One can fight a danger only when one is armed with solid facts and spurred on by an unwavering faith and determination.

(*The Autobiography of Eleanor Roosevelt*)

One must be friends with oneself before one can achieve contentment and courage, and one must understand that living is an adventure—an adventure that can only be savored by the courageous. There is, of course, no such thing as real security for anyone in the world. There may be sorrow and suffering around the corner for anyone at any time. Nevertheless, life is worth living and it is worthwhile to love, even though that very love may bring you suffering.

(My Day, October 2, 1946)

If you face life with a spirit of adventure and with courage you will get more out of it than if you are timid and unimaginative.

(If You Ask Me, *McCall's*, December 1952)

Unless people are willing to face the unfamiliar they cannot be creative in any sense, for creativity always means the doing of the unfamiliar, the breaking of new ground.

(*Tomorrow Is Now*)

WORRY

Many of our worries lie in anticipation of things which may never happen.

(If You Ask Me,
Ladies' Home Journal, June 1944)

It is foolish to worry, for all of us know that whatever comes we have to meet it.

(My Day, August 16, 1941)

I do not believe in worrying about something I can do nothing about.

(My Day, December 18, 1951)

The best way to alleviate worry is to do all you can.

(If You Ask Me, *McCall's*, July 1950)

CRITICISM

People are very apt to grow critical and excited when they know very little about a situation.

(My Day, April 18, 1946)

Destructive criticism is always valueless and anyone with common sense soon becomes completely indifferent to it.

("How to Take Criticism,"
Ladies Home Journal, November 1944)

Curiously enough, it is often the people who refuse to assume any responsibility who are apt to be the sharpest critics of those who do.

(*You Learn by Living*)

It is perfectly natural that there should be people who like neither me nor my ideas nor my husband and what he stands for in public life. . . .

(If You Ask Me,
Ladies' Home Journal, May 1943)

I have always kept a file of disagreeable letters, which I call my "hair shirt" file, so that in case at any time I should feel the nice letters were giving me too much satisfaction I would always have something to turn to to keep my feet reasonably safely on the ground.

(If You Ask Me, *McCall's,* January 1950)

LIVING WITH OTHERS

There is very little actual fundamental law. Really only "Love one another." The rest is all interpretation.

(My Day, February 13, 1937)

Whether we recognize it or not, no home is an isolated object. We may not recognize it, and we may try to narrow ourselves so that our interest only extends to our immediate home circle, but if we have any understanding at all of what goes on around us, we soon see how outside influences affect our own existence.

("In Defense of Curiosity," *Saturday Evening Post,*
August, 24, 1935)

Learning about other people does not mean that, of necessity, one must emulate them. It does mean one will have a better understanding of their motives and reactions.

(My Day, February 11, 1943)

You can never stop people from talking but you do not have to believe everything that you hear.

(Letter to L. M. Deemer, August 29, 1949)

Every impulse to draw people closer together is worth thinking about. . . .

(My Day, January 13, 1950)

Human beings are so made that they cannot help disagreeing with each other, now on one subject and now on another.

(My Day, July 19, 1946)

We are shocked and surprised when we find that other people doubt our intentions, forgetting entirely that their background is not like ours.

(My Day, August 12, 1946)

When you play on a team, you accept certain things as a member of that team. If certain objectives have been decided upon, you take direction to achieve them.

(My Day, March 26, 1946)

Caring comes from being able to put yourself in the position of the other person. If you cannot imagine, "This might happen to me," you are able to say to yourself with indifference, "Who cares?"

(*The Autobiography of Eleanor Roosevelt*)

Each of us carries some little influence in our own particular area.

(My Day, April 22, 1957)

I still believe it is the personal and individual thing which is done by one human being for another that is the most valuable assistance which can be rendered in time of trouble.

(My Day, January 4, 1938)

We never seem to take into consideration that there is an art in human relationships and that our appreciation of people may vary in just the same way as does our appreciation of art.

(My Day, January 2, 1937)

RESPONSIBILITY

As a matter of fact most people want to be told, they don't want to decide for themselves. They like to be able to blame someone afterwards if they make a wrong decision!

(Letter to Joseph P. Lash, July 20, 1944)

It takes honesty and courage to accept the full responsibility when your first choice has been wrong; it takes honesty and courage to acknowledge that the fault was yours and you have no excuses to make.

(*You Learn by Living*)

If you live long enough and carry responsibility long enough, you do build up sufficient confidence so that at times people may follow even when they are frankly told that it is not possible to give them the reasons for a decision.

(My Day, March 13, 1950)

I think older people have a responsibility to help young people to assume more and more of the management of the world in which they live.

(My Day, October 16, 1946)

CHOICES

We all create the person we become by our choices as we go through life. In a very real sense, by the time we are adult, *we are the sum total of the choices we have made.*

(*You Learn by Living*)

In the long run, we shape our lives and we shape ourselves. The process never ends until we die. And the choices we make are ultimately our own responsibility.

(You Learn by Living)

Words and promises are never enough. Performance is what counts.

(My Day, October 15, 1946)

One's philosophy is not best expressed in words; it is expressed in the choices one makes.

(You Learn by Living)

COMPROMISE

Negotiation and compromise are very good words but you have to know what you compromise and with whom you are negotiating.

(Letter to Sara Gould, January 19, 1951)

You may have to retreat a little here, and you accept what seems to be the inevitable there, but all the time you move forward on your broad general front.

(My Day, February 13, 1951)

Just as all living is adjustment and readjustment, so all choice, to some extent, must be compromise between reality and a dream of perfection.

(You Learn by Living)

ACCEPTANCE

It is rather useless to talk about something you can't do anything to change.

(My Day, June 21, 1957)

You have to take things as you find them in this world and do what you can to remedy them.

(My Day, June 27, 1946)

With my enemies I am never inclined to argue. What is the use?

(My Day, August 26, 1946)

MANAGING CHANGE

Seize on every thing that comes your way which makes life more interesting, or more agreeable, meet whatever circumstances arise in what some critics will call a haphazard or opportunist manner, but in what you yourself may consider is the only way to face an ever changing life in an ever changing world.

(My Day, January 1, 1937)

Often people have asked me, "How do you recover from disaster?" I don't know any answer except the obvious one: You do it by meeting it and going on. From each you learn something, from each you acquire additional strength and confidence in yourself to meet the next one when it comes.

(*You Learn by Living*)

How hard it is for human beings to learn that the only safety there is, lies in being prepared for any eventuality.

(My Day, December 9, 1941)

All intelligent people change their minds in view of changed circumstances and conditions. Only stupid people remain rigid and inflexible in their opinions and ideas.

(My Day, October 25, 1945)

GIVING AND RECEIVING

When giving is too easy or is purely a matter of giving material things, it ceases to bring real happiness.

(*It's Up to the Women*)

If you are doing something that exploits people and is purely selfish they will always know it. But if you act from a real concern for people, a desire to help them to develop themselves, you will find warm cooperation and eventually an ability to work together that is fruitful because it is voluntary and not imposed. But it must be a natural growth based on mutual respect and trust.

(*Tomorrow Is Now*)

We must learn never to demand of someone else what is not freely offered us. This can apply to one's husband or wife, to one's children, particularly after they have left home, to one's friends. What is freely given in love or affection or companionship one should rightly rejoice in. But what is withheld one must not demand.

(*You Learn by Living*)

It is an art to receive graciously and to give thanks and appreciation with a warm heart and no feeling of hesitation because we have given nothing.

(*My Day*, December 25, 1937)

You must of course, give to the extent of your ability but there is an art also in receiving.

(*My Day*, May 11, 1937)

I think I'm pretty much of a fatalist. You have to accept whatever comes and the only important thing is that you meet it with courage and with the best that you have to give.

(*This I Believe*, 203)

SUCCESS

Success must include two things: the development of an individual to his utmost potentiality and a contribution of some kind to one's world.

(*You Learn by Living*)

A successful life for a man or for a woman seems to me to lie in the knowledge that one has developed to the limit the capacities with which one was endowed; that one has contributed something constructive to family and friends and to a home community; that one has brought happiness wherever it was possible; that one has earned one's way in the world, has kept some friends and need not be ashamed to face oneself honestly.

(If You Ask Me,
Ladies' Home Journal, November 1941)

I am inclined to think that being a success is tied up very closely with being one's own kind of individual.

(*You Learn by Living*)

To leave the world richer—that is the ultimate success.

(*You Learn by Living*)

FAME

The more you live in a "gold-fish bowl," the less people really know about you!

(My Day, January 7, 1936)

People can gradually be brought to understand that an individual, even if she is a President's wife, may have independent views and must be allowed the expression of an opinion. But actual participation in the work of the Government, we are not yet able to accept.

(My Day, February 23, 1942)

For many years of my life I realized that what my husband was attempting to do was far more important than anything which I could possibly accomplish; and therefore I never said anything, or wrote anything, without first balancing it against the objectives which I thought he was working for at the time. We did not always agree as to methods, but our ultimate objectives were fortunately very much the same.

("Why I Do Not Choose to Run,"
Look, July 9, 1946)

I find if I never look at anyone and move fast I am frequently not recognized.

(If You Ask Me, McCall's, January 1953)

I have never felt famous.

(If You Ask Me, McCall's, May 1954)

AGING

The greatest tragedy of old age is the tendency for the old to feel unneeded, unwanted, and of no use to anyone; the secret of happiness in the declining years is to remain interested in life, as active as possible, useful to others, busy, and forward looking.

(Book of Common Sense Etiquette)

We should think of more ways in which we can use our retired people and the mature people who come to this country. We should give some recognition to the skills they can bring us. . . . We are allowing waste in our most valuable material, our human resources.

(You Learn by Living)

It is a mistake, I think, for older people to move in with a young family. . . . [M]odern life is lived in smaller quarters, it is more hurried and less leisurely, and I think it is

harder for different generations to adjust to each other.
Personally, therefore, I should prefer to try to find a
home in which I felt I could have some privacy. . . .

(If You Ask Me, *McCall's,* May 1955)

With advancing age, you accept the blows of life more
philosophically. In fact, you accept life, which is perhaps nature's way of preparing you to accept death.

(*You Learn by Living*)

One of the blessings of age is to learn not to part on a
note of sharpness, to treasure the moments spent with
those we love, and to make them whenever possible
good to remember, for time is short.

(My Day, February 5, 1943)

As you grow older, sometimes you have the good for-
tune to outgrow some of the misunderstandings of your
youth and to learn the real values of people.

(My Day, January 9, 1957)

One of the penalties of growing older is that you become more and more conscious that nothing in life is
very permanent.

(My Day, October 9, 1937)

I could not, at any age, really be contented to take my
place in a warm corner by the fire and simply look on.

(*The Autobiography of Eleanor Roosevelt*)

Nine

EMOTIONS AND HABITS

FOR SOMEONE WHO WAS NOT KNOWN FOR BEING demonstrative, Eleanor Roosevelt wrote a great deal about emotion. She was particularly concerned about the link between emotion and behavior. For example, she considered fear to be the "great crippler" and inhibitor of action. Courage, on the other hand, was empowering and "in the long run . . . easier." Hope was a more "intelligent" response to problems than cynicism, which she considered "a form of philosophical defeat."[1]

These were not the pious platitudes of someone who had never struggled. On the contrary, Eleanor spent years learning to harness her emotions and make them work for her instead of against her. Her chief tools in that effort were self-discipline and the willingness to face her fears and insecurities, analyze them, "and, in the end, act." Often her actions took the form of doing whatever needed to be done in a given situation. For example, she conquered her fear of public speaking by making speeches and overcame her fear of insanity by visiting mentally ill veterans during World War I. Confronted by her husband's romance with another woman, she made the decision to stay in her marriage while pursuing an independent life and career. Working for causes she believed in built her confidence even when she suffered setbacks, as she often did when dealing with issues like racial discrimination. Regardless of whether she won or lost, she persisted in her quest to vanquish fear until she reached the point where she could truthfully say she feared no person or circumstance.[2]

ANGER

In our family we have always said there are two kinds of anger. One kind you have to fight and conquer, since it shows weakness. The other kind is righteous indignation, which can be indulged in with a clear conscience.

(My Day, July 29, 1946)

Righteous indignation may be valuable at times, but it must be used by those who have thought out very carefully the end results they want to achieve.

(My Day, January 29, 1951)

The only harm that comes from self-indulgence in bitterness and anger is what happens to the person so indulging herself.

(My Day, September 19, 1942)

CYNICISM

Cynicism seems to me a form of philosophical defeat. It comes only when you have given up any thought or hope of achievement.

(*You Learn by Living*)

Sometimes I wonder if we believed all the pessimistic things that we hear whether we would ever have the courage to go ahead and try anything new.

(My Day, January 12, 1937)

FEAR

Fear is a bad thing at all times and should be eliminated from our lives as much as possible.

(*It's Up to the Women*)

Fear has always seemed to me to be the worst stumbling block which anyone has to face. It is the great crippler.

(*You Learn by Living*)

In this world, however, most of us are motivated by fear—governments more, perhaps, even than individuals.

(My Day, February 10, 1960)

My greatest fear has always been that I would be afraid—afraid physically or mentally or morally—and allow myself to be influenced by fear instead of by my honest convictions.

(If You Ask Me,
Ladies' Home Journal, August 1942)

Our dread of meeting some new situation is really a fear left over from our childhood when everything new we tried was an unknown adventure.

(*It's Up to the Women*)

We should try above everything else to keep away from all children any sense of fear, in these early years. Fear of their teachers, fear of the ridicule of their contemporaries, fear of their own inability to meet whatever situations they may have to face in life.

(*It's Up to the Women*)

No one from the beginning of time has ever had security. When you leave your house you do not know what will happen on the other side of the door. Anything is possible. But we do not stay home on that account. After all, the man who cowers under a tree in a storm, thinking that he is secure, merely runs more risk from the lightning.

(*Tomorrow Is Now*)

You cannot be protected from a person who does not care whether he is caught or not. The only possible course is to put the thought of danger out of your mind and go ahead with your job as you feel you must, regardless of what might be called its occupational risks.

(*The Autobiography of Eleanor Roosevelt*)

The discipline one imposes on oneself is the only sure bulwark one has against fear.

(*You Learn by Living*)

Having learned to stare down fear, I long ago reached the point where there is no living person whom I fear, and few challenges that I am not willing to face.

(*The Autobiography of Eleanor Roosevelt*)

No matter how hard hit you are, you can face what has to be faced if you have learned to master your own fears.

(*You Learn by Living*)

We are at present in the grip of a wave of fear which threatens to overcome us. I think we need a rude awakening, to make us exert all the strength we have to face facts as they are in our country and in the world, and to make us willing to sacrifice all that we have from the material standpoint in order that freedom and democracy may not perish from this earth.

("Keepers of Democracy,"
Virginia Quarterly Review, January 1939)

I haven't ever believed that anything supported by fear can stand against freedom from fear. Surely we cannot be so stupid as to let ourselves become shackled by senseless fears. The result of that would be to have a system of fear imposed on us.

(*You Learn by Living*)

It is fear of each other that makes us do so many so-called aggressive things. To show that we are not afraid we both boast of our might and of our power. We hold weapons and the threat of new inventions somewhere close at hand.

(My Day, January 4, 1950)

Once fear crops up between groups of people within a community, there is little hope of any real understanding and confidence developing between them.

(My Day, July 9, 1946)

You fear in apprehension far more than you actually suffer in reality.

(*You Learn by Living*)

People who "view with alarm" never build anything.

(*Tomorrow Is Now*)

COURAGE

The answer to fear is not to cower and hide; it is not to surrender feebly without contest. The answer is to stand and face it boldly. Look at it, analyze it, and, in the end, act.

With action confidence grows.

(*Tomorrow Is Now*)

The danger lies in refusing to face the fear, in not daring to come to grips with it. If you fail anywhere along the line, it will take away your confidence. . . . *You must do the thing you think you cannot do.*

(*You Learn by Living*)

The encouraging thing is that every time you meet a situation, though you may think at the time it is an impossibility and you go through the tortures of the damned, once you have met it and lived through it you find that forever after you are freer than you ever were before. If you can live through that you can live through anything. You gain strength, courage, and confidence by every experience in which you really stop to look fear in the face.

(*You Learn by Living*)

In the long run there is no more liberating, no more exhilarating experience than to determine one's position, state it bravely, and then *act boldly*. Action brings with it its own courage, its own energy, a growth of self-confidence that can be acquired in no other way.

(*Tomorrow Is Now*)

I think there is a kinship among all brave souls—those who have died bravely and those who are living bravely.

(My Day, March 3, 1950)

Courage is more exhilarating than fear and in the long run it is easier. We do not have to become heroes overnight. Just a step at a time, meeting each thing that comes up, seeing it is not as dreadful as it appeared, discovering we have the strength to stare it down.

(*You Learn by Living*)

It is not only physical courage which we need, the kind of physical courage which in the face of danger can at least control the outward evidences of fear. It is moral courage as well, the courage which can make up its mind whether it thinks something is right or wrong, make a material or personal sacrifice if necessary, and take the consequences which may come.

("Keepers of Democracy,"
Virginia Quarterly Review, January 1939)

Different days have different difficulties, and those that we face today are perhaps not as simple as those which faced the pioneers. The solutions are more difficult to find but the same kind of courage and determination which conquered their difficulties, applied to ours will do the trick, I think.

(*My Day*, February 8, 1937)

SELF-DISCIPLINE

Merely listening to so many points of view is not only good discipline but opens up vast fields of knowledge

which most of us need to explore if we are going to meet
the needs of the next few years intelligently.

(My Day, February 14, 1946)

Self-discipline brings its own reward.

(My Day, March 4, 1946)

Anyone who does not have enough self-control to live
within the bonds of moderation is a slave in the very
truest sense of the word.

(If You Ask Me,
Ladies' Home Journal, June 1944)

SELF-CONFIDENCE

To do anything constructive or creative in this world,
people must have some self-confidence. Therefore
people who love them must always be careful even
in giving their honest criticism and opinions, not to
destroy completely an individual's faith in his own
judgment!

("How to Take Criticism,"
Ladies' Home Journal, November 1944)

It's always easy to blurt out all you know, to try to get
your burden shared by other people. It's far more

difficult to take the best advice you can get, make your own decisions, knowing that you will only be adding to the risks of the situation if you try to turn the decision over to others who cannot have the same background and knowledge.

(*Over Our Coffee Cups*, December 7, 1941)

There is nothing more exciting in the world than to be conscious of inwardly achieving something new.

(*The Moral Basis of Democracy*)

No one can expect always to deal correctly with every question, but a confident approach gives one a better chance of success.

(My Day, February 19, 1946)

No man is defeated without until he has first been defeated within.

(*You Learn by Living*)

HOPE

Men who lack vision are poor in hope. They turn their backs on the future and live in the past.

(Address to the Democratic
National Convention, July 22, 1952)

The hope of the world lies in acceptance of a philosophy which has come down to us through the ages. Love
can be stronger than hate, but we as individuals have
to see to it that love and not hate is the basis of our
action.

(My Day, February 2, 1946)

Surely, in the light of history, it is more intelligent to
hope rather than to fear, to try rather than not to try.
For one thing we know beyond all doubt: Nothing has
ever been achieved by the person who says, "It can't be
done."

(*You Learn by Living*)

HAPPINESS

There are three fundamentals for human happiness—
work which will produce at least a minimum of material
security; love and faith. These things must be made possible for all human beings, men and women alike.

(My Day, January 31, 1936)

Happiness is not a goal, it is a by-product. Paradoxically,
the one sure way not to be happy is deliberately to map
out a way of life in which one would please oneself completely and exclusively. After a short time, a very short
time, there would be little that one really enjoyed. For

what keeps our interest in life and makes us look forward to tomorrow is giving pleasure to other people.

(*You Learn by Living*)

Someone once asked me what I regarded as the three most important requirements for happiness. My answer was: "A feeling that you have been honest with yourself and those around you; a feeling that you have done the best you could both in your personal life and in your work; and the ability to love others."

But there is another basic requirement . . . that is the feeling that you are, in some way, useful.

(*You Learn by Living*)

As I grow older I realize that the only pleasure I have in anything is to share it with someone else. That is true of memories, and it is true of all you do after you reach a certain age. The real joy in things, or in the doing of things, just for the sake of doing or possessing, is gone; but to me the joy in sharing something that you like with someone else is doubly enhanced.

("What I Hope to Leave Behind,"
Pictorial Review, April 1933)

Enjoy everything that gives you pleasure and make no comparisons!

(*My Day*, April 24, 1937)

When we admit that one can not retrace one's steps and live life over again, we may accept the fact that we might as well savor it and enjoy it as we go along and not always be striving for something in the future which after all we may never achieve nor enjoy.

(*It's Up to the Women*)

TRANQUILITY

In this restless age, even if it must be done by artificial means, it is necessary to acquire an atmosphere of peace for part of every day. It is necessary not only to one's own health but also to the happiness of those in the home.

(*It's Up to the Women*)

Repose is not a question of sitting still. It is a kind of spiritual attitude; no superficial human being can have it; real repose requires depth, a rich personality. The person possessing it can create a feeling that life flows smoothly and peacefully. Though they may never sit with folded hands, you may be able to sit with them and experience complete relaxation.

("In Defense of Curiosity,"
Saturday Evening Post, August 24, 1935)

Ten

WAR AND PEACE

ELEANOR ROOSEVELT WAS AN ARDENT CAMPAIGNER for peace beginning in the 1920s with her work organizing the Bok Peace Prize. In the 1930s she worked with several women's peace organizations, including the National Committee on the Cause and Cure of War, the Women's International League for Peace and Freedom, and the American Friends Service Committee. Devoted as she was to the cause, Eleanor was no pacifist. She understood the need to fight World War II, and during the Cold War she favored a strong military. However, she never reconciled herself to the carnage of war, considering it "criminal stupidity." The advent of the nuclear era only hardened her determination to fight for peace.[1]

It was a conflict she fought with vigor and determination for the rest of her life. Eleanor understood that making peace was no pie-in-the-sky endeavor. It required tenacious effort on the part of both individuals and groups—the same kind of effort that went into fighting a war. Effective peacemaking also required a commitment to justice, because "only justice keeps the people of the world at peace." (That insight lay at the root of her work on the Universal Declaration of Human Rights, which, under her leadership, set an international standard for equality and justice.) Perhaps most important, she recognized that the work of peace was continuous and unending. Yet in the nuclear age she believed it was the only rational choice.[2]

WAR

All people desire peace, but they are led to war because what is offered them in this world seems to be unjust, and they are constantly seeking a way to right that injustice.

(*The Moral Basis of Democracy*)

If life is not worthwhile, war is thought to be no worse. . . .

(My Day, May 30, 1957)

War is the result of spiritual poverty. People say that war is the cause of a great many of our troubles; but in the first analysis it is the fact that human beings have not developed the ability to rise above purely selfish interest which brings about war. Then war intensifies all of our social problems and leaves us groping for the answers.

(*The Moral Basis of Democracy*)

War is the most costly thing that can befall us. It means complete destruction from which there is no return.

(My Day, May 22, 1957)

Someone may say: "But we need only to go on until the men who at present have power in the world and who believe in force are gone." But when in the past has there been a time when such men did not exist? If our civilization is to survive and democracies are to live, then the people of the world as a whole must be stronger than such leaders.

("Keepers of Democracy,"
Virginia Quarterly Review, January 1939)

What a fearful waste it is that we have to go killing each other before even a difference of opinion can be settled amongst people of the same nation. I think the part which annoys one is the stupidity of wasting so much energy, vitality and ability just to destroy people and things when at the moment everywhere we need to conserve people to build up material possessions.

(My Day, September 3, 1936)

It seems insanity to me to try to settle the difficult problems of today by the unsatisfactory method of going to war. If you kill half the youth of the continent, the problems will be no nearer to solution, but the human race will be that much poorer.

(My Day, September 17, 1938)

When we go out to kill each other instead of to help each other, we give valid proof of the failure of our type of civilization.

(My Day, March 29, 1946)

War no longer deals with soldiers alone, it deals just as harshly with men, women and children and that is why, if our civilization is to continue, war must come to an end.

(My Day, January 7, 1946)

Of course, when we talk of "the front" in connection with future wars, we are taking it for granted that future wars will be much like those of the past, whereas most people believe that future wars will have no fronts.

(*This Troubled World*)

I wish that those who are belligerent today could realize that all wars eventually act as boomerangs and the victor suffers as much as the vanquished.

(My Day, February 7, 1939)

Nobody wins a modern war.

("Liberals in This Year of Decision,"
Christian Register, June 1948)

POSTWAR

The period after a war is very hard. The incentive to sacrifice is over, and yet many of the problems are as difficult, and sometimes more difficult than if the war were still on.

(My Day, January 23, 1946)

Armies of occupation cannot take the place of a country's own civilian leaders.

(My Day, February 14, 1946)

Any occupation force, whether good or bad, just or unjust, is detested by the people it rules, and I have never had reason to believe that an exception is made of American troops.

(*The Autobiography of Eleanor Roosevelt*)

A war ends and there is a shuffle in the cards of power politics. The man who was our enemy is now our friend; the friend at whose side we fought so gladly and so proudly has now become the enemy. From a long-range viewpoint all this appears to be nothing but criminal stupidity.

(*Tomorrow Is Now*)

NUCLEAR WAR

Once a weapon is discovered, it will always be used by those who are in desperate straits.

(My Day, September 25, 1945)

The day we found the secret of the atomic bomb, we closed one phase of civilization and entered upon another.

(My Day, September 25, 1945)

We have only two alternative choices: destruction and death—or construction and life!

(My Day, August 8, 1945)

We are not aggressive; we do not want to attack any nation. What we want is to be strong enough so that no nation will be tempted to attack us.

(My Day, January 30, 1960)

The armament race and rivalry in scientific research and in the economic field will mean that the cost of government in the various countries will mount steadily. Instead of having money for constructive purposes such as education, better health, better housing, and more social security for the average man and woman, they will have to pay the cost of the new fears that stalk the world.

(My Day, March 16, 1946)

No one nation can be arrogant enough to believe that if they build up a tremendous war machine they will not have the same temptation to use it as other nations have in the past. We, like other nations, must have adequate forces for defense. Of course, this does not mean that any one nation, by itself, can achieve world peace by being the world's policeman.

(Interview with Milton Cross, January 9, 1938)

The only eventual security against nuclear weapons is disarmament. . . . A single great power left outside a disarmament agreement could determine the course of the world if it wished.

(*Tomorrow Is Now*)

THE UNITED NATIONS

On the success or failure of the United Nations Organization may depend the preservation and continuance of our civilization.

(My Day, January 5, 1946)

Today, every human being in the world stands in constant peril from irresponsible use of nuclear power. But today, also, we have created the only machinery for peace that has ever functioned. That, of course, is the United Nations. . . .

Because it is the work of men and women, of fallible human beings, the United Nations is not a perfect instrument. *But it is all we have.*

(*Tomorrow Is Now*)

The United Nations is not a club of congenial people; it is, as it should be, a reflection of the whole world, with its turmoil, its conflicting interests, its diverse viewpoints.

(*Tomorrow Is Now*)

The United Nations is not a cure-all. It is only an instrument capable of effective action when its members have a will to make it work.

(Address to the Democratic National
Convention, July 22, 1952)

The United Nations was not organized to make peace. It was organized to help maintain peace in the world where as yet the warring nations have not made peace.

(Letter to Margaret Carberry, February 22, 1949)

With all our agitation about peace, we lose sight of the fact that with the proper machinery it is easier to keep out of situations which lead to war than it is to bring about peace once war is actually going on.

(*This Troubled World*)

One great strength of the United Nations is often not recognized; indeed, it is often regarded as a weakness. That is the amount of talk that goes on. Now the value of a public forum where people can protest their wrongs is enormous. In the first place, they are able to bring their problems and their complaints before world opinion; to arouse wide discussion about how their problems can be solved. But the second advantage is that talk is a wonderful way of letting off steam. It is a kind of safety valve. As long as men are arguing about the situation in words, they are not trying to solve it with bullets.

(*Tomorrow Is Now*)

As long as the United Nations is really united, there is a bridge which the peoples of the world can use to reach each other; and little by little, I believe, they will increase their understanding and their confidence in each other. Once we are divided into two camps, and armed camps at that, the future will be bleak indeed.

(My Day, May 1, 1950)

There is no question, for instance, that ultimately we will all have more security if we have a greater sense of interdependence, put more strength in the United Nations, and count less on our individual strengths.

(If You Ask Me, *Ladies' Home Journal,* January 1947)

Without the United Nations our country would walk alone, ruled by fear instead of confidence and hope. To weaken or hamstring the United Nations now, through lack of faith and lack of vision, would be to condemn ourselves to endless struggle for survival in a jungle world.

(Address to the Democratic National Convention, July 22, 1952)

THE UNIVERSAL DECLARATION OF HUMAN RIGHTS

[The Universal Declaration of Human Rights] is not a treaty; it is not an international agreement. It is not and does not purport to be a statement of law or of legal obligation. It is a declaration of basic principles of human rights and freedoms . . . to serve as a common standard of achievement for all peoples of all nations.

("General Assembly Adopts Declaration of Human Rights,"
Department of State Bulletin, December 19, 1948)

When all is said and done, the declaration will set a standard for human rights and freedoms, and if these standards are recognized as good I believe peoples throughout the world, who feel they are not being treated fairly, will gain a knowledge of the declaration. Then that silent pressure of the masses will be felt in the Kremlin in Moscow or any other government abode the world over.

(My Day, December 10, 1948)

The future must see the broadening of human rights throughout the world. People who have glimpsed freedom will never be content until they have secured it for themselves. In a true sense, human rights are a fundamental object of law and government in a just society. Human rights exist to the degree that they are respected by people in relations with each other and by governments in relations with their citizens.

("The Struggle for Human Rights,"
September 28, 1948)

Abuses anywhere, however isolated they may appear, can end by becoming abuses everywhere.

(*The Autobiography of Eleanor Roosevelt*)

Where, after all, do universal human rights begin? In small places, close to home—so close and so small that they cannot be seen on any maps of the world. Yet they *are* the world of the individual person; the neighborhood he lives in; the school or college he attends; the factory, farm, or office where he works. Such are the places where every man, woman, and child seeks equal justice, equal opportunity, equal dignity without discrimination. Unless these rights have meaning there, they have little meaning anywhere. Without concerted citizen action to uphold them close to home, we shall look in vain for progress in the larger world.

("Where Do Human Rights Begin?,"
remarks at the United Nations, March 27, 1958)

Human rights really exist only in the framework of democracy and freedom.

("The Elementary Teacher as a Champion of Human Rights,"
The Instructor, September 1951)

PEACE

At the time of World War I, I felt keenly that I wanted to do everything possible to prevent future war, but I never felt it in the same way that I did during World War II. During this second war period I identified myself with all the other women who were going through the same slow death, and I kept praying that I might be able to prevent a repetition of the stupidity called war.

I have tried, ever since, in everything I have done, to keep that promise I made to myself, but the progress that the world is making toward peace seems like the crawling of a little child, halting and slow.

(*The Autobiography of Eleanor Roosevelt*)

Negotiation, mediation or arbitration are just words, but any one of them if put into practice now by people who really want to keep peace, might mean life instead of death to hundreds of thousands of young men.

(My Day, August 26, 1939)

Peace is no longer a question of something we hope to attain in the future. It is an absolutely vital necessity to the continuation of our civilization on earth.

(My Day, October 6, 1945)

It must be the combined voice of the world as a whole insisting that because of fear of one another we refuse to continue going down the path to destruction.

(My Day, February 16, 1950)

The greed, suspicion and fear which have created wars in the past will create them again unless, through education and understanding, human beings can be brought to see that their own best interests lie along new lines of development. If we hope to prosper, others must prosper too, and if we hope to be trusted, we must trust others.

(My Day, January 5, 1946)

We are engaged in the war for peace in which there enter questions of world economy, food, religion, education, health and social conditions, as well as military and power conditions.

(Memo for Harry S. Truman, December 28, 1948)

Since European nations are more international-minded, they are not apt to forget that peace requires as much attention as war. But the United States, because of its early isolation, has lived in what many might call a Fools' Paradise.

(My Day, February 12, 1946)

We will have to want peace, want it enough to pay for it, pay for it in our own behavior and in material ways. We will have to want it enough to overcome our lethargy and go out and find all those in other countries who want it as much as we do.

(*This Troubled World*)

To achieve peace we must recognize the historic truth that we can no longer live apart from the rest of the world. We must also recognize the fact that peace, like freedom, is not won once and for all. It is fought for

daily, in many small acts, and is the result of many individual efforts.

(Address to the Democratic National
Convention, July 22, 1952)

There is no point definitely set when we win the peace.
It goes on, that effort, day in and day out, all through
our lives, perhaps through the lives of our children
and grandchildren. But every year . . . that we keep the
peace, that means one step more towards our goal.

(Speech at the Phi Beta Kappa Association
Founders' Day Dinner, February 25, 1949)

Peace is something we want to work for, day in and day
out, but we want to work for it with the knowledge that
only justice keeps the people of the world at peace.

(My Day, December 9, 1942)

It takes just as much determination to work for peace as
it does to win a war.

(Address to Roosevelt College, November 16, 1945)

Peace, like freedom, is elusive, hard to come by, harder
to keep. It cannot be put in a purse or a hip pocket and
buttoned there to stay.

(Address to the Democratic National
Convention, July 22, 1952)

Peace without freedom is a stagnant pool. It may look
alluring at a distance, the lily pads may gleam white in

the sun, but underneath the water is foul. Freedom is a fresh and running stream in which there is refreshment for the soul—a thing of beauty and power to be used for the common good.

(Statement for Paris, April 14, 1949)

We believe in one actual way to peace—making a fundamental change in human nature. Over and over again people will tell you that that is impossible. I cannot see why it should be impossible when the record of history shows so many changes already gone through. . . .

(*This Troubled World*)

Perhaps peace in the world starts in one's own heart and will become a reality only when innumerable people throughout the world find peace within themselves.

(My Day, March 19, 1957)

If peace is going to come about in the world, the way to start is by getting a better understanding between individuals. From this germ a better understanding between groups of people will grow.

(*This Troubled World*)

If peace depends upon us—and many feel it must— then we will achieve it only by giving leadership. We will achieve it only by making sacrifices. We cannot tell other people what to do. We must show them by our example what we think is right, and that will lead them to recognize their own responsibilities.

(My Day, November 14, 1946)

Perhaps the truth of it is that we must want to live in peace at home before we can translate this desire for peace in the world into practice.

(My Day, March 19, 1957)

Each of our country's citizens should realize that it is necessary for him to learn how to get along with people of different races and nationalities at home and to spread this spirit to international relations. This responsibility no longer can be left in the hands of a few. The business of keeping out of war is the business of every citizen in a democracy.

(My Day, May 30, 1957)

Somehow we must learn to live together or at some point we are apt to die together.

(My Day, January 22, 1959)

Eleven

THE UNITED STATES AND THE WORLD

 AMERICA'S ROLE IN THE WORLD BECAME ONE OF ELEANOR Roosevelt's central preoccupations after World War II. As one of the first U.S. delegates to the United Nations, she was well placed to judge how other nations saw America—and the view was not always flattering: "No one likes the rich uncle who flaunts his wealth in the face of your poverty; who will help you perhaps—but on his own terms . . . ," she wrote in her 1953 book *India and the Awakening East*. "This . . . is not a fair description of our attitude: but, nevertheless, fair or not, it is the way many people see us."[1] Americans' perceptions of other nations were often skewed as well. At best, "we in

the United States . . . tend to believe that the conditions
of all nations, and the desires of all peoples coincide
with our own." At worst, Americans think "we do not
have to worry about understanding other people's prejudices and customs, since we can go our own way. With
hard work, we expect to get along regardless of what
happens to others."[2]

Neither attitude, she believed, served American interests in the Cold War era. Instead she urged her fellow
Americans to accept the responsibilities of international
leadership, saying that "the world is waiting for us to provide an example of dynamic drive, a bold reaffirmation of
the values on which our nation was founded." Rather than
retreat into isolationism or rely on a foreign policy that
prioritized military foreign aid as a way to make friends
abroad, she challenged her contemporaries to spend their
time, talent, and treasure learning about the needs and
aspirations of other countries and working with them to
help them reach their goals.[3]

At a time when new nations were emerging from
old colonial empires, she understood that the successful
spread of American values hinged on mutual understanding and trust. "We must relearn the meaning of that noble
word *respect*," she wrote. "That is the only sound and enduring basis for any relationship among peoples, as it is
among individuals."[4]

INTERNATIONAL RELATIONS

The human family, no matter where it lives or what language it speaks, is closely tied together and . . . somehow we must learn to understand each other's pain.

(My Day, May 17, 1957)

Difficulties between people or nations that are not talked out create a potential bitter feeling of injustice on all sides.

(My Day, February 8, 1946)

Human beings have a breaking point if denied an outlet for their emotions and convictions. Then violence may seem to be the only answer, and that hurts us, both at home and abroad.

(My Day, May 24, 1957)

There is no real reason for thinking we are living in a peaceful world.

(My Day, February 4, 1960)

The great trouble internationally today is that we have built no confidence in each other. We think in the same old terms of individual strength and control through our own power.

(My Day, May 7, 1946)

Security requires both control of the use of force and the elimination of want. No people are secure unless they have the things needed not only to preserve existence, but to make life worth living.

(*My Day, January 5, 1946*)

Lack of integrity, or perhaps we should call it more politely the desire to be a little more clever than one's neighbor, is what promotes a constant attitude of suspicion amongst nations. This will exist until we have accomplished a change in human nature.

(*This Troubled World*)

Differences of opinion arise in the family as to conduct or as to likes and dislikes. Why should we expect therefore, that nations will not have these same differences and quarrels? Why do we concentrate on urging them not to have any differences?

(*This Troubled World*)

Most of us are taught as children to count to thirty before we opened our mouths when we were angry, and that same lesson should apply to nations.

(*My Day, March 18, 1936*)

It seems to me now that we should turn to the countries that are willing to cooperate with us, and not behave like white rabbits when we deal with those who seem to scorn us.

(*My Day, January 8, 1957*)

Unless we have freedom of information, there will be no knowledge of whether nations do or do not carry out their promises. . . .

(My Day, June 5, 1946)

Day by day we must be reminded that our world is one world.

(My Day, January 3, 1946)

The world cannot be understood from a single point of view.

(*Tomorrow Is Now*)

AMERICAN VALUES

We seem to have forgotten to weigh our values and to realize that we have to pay for the things we want. The payment which can bring about friendly and peaceful solutions is infinitely less costly than the payments which will have to be made if we are going to be an enemy to all the world.

(My Day, August 23, 1946)

Only time will prove whether we can live up to the high standards we have set for ourselves.

(My Day, April 20, 1950)

THE UNITED STATES' VIEW OF THE WORLD

We are, fortunately, situated with enough land for our needs at present and an amazing amount of natural resources. We need comparatively little from other nations. We have few traditional fears. We are a self-confident and self-reliant people, and on the whole self-sufficient. Our fortunate situation, however, does not mean that we can persuade others to have the same sense of security that we have ourselves. And a sense of security is vital to a peaceful world.

(Interview with Milton Cross, January 9, 1938)

It is a traditional American feeling that we do not have to worry about understanding other people's prejudices and customs, since we can go our own way. With hard work, we expect to get along regardless of what happens to others.

(My Day, September 28, 1945)

We in the United States tend sometimes to think of international questions as they concern our own particular problems and we tend to believe that the conditions of all nations, and the desires of all peoples coincide with our own. This is something we must remember to check . . . for different conditions bring different conceptions of what rights and freedoms should be.

(My Day, January 7, 1950)

So often in the United States we take so much for granted that we find it hard to explain to anyone else what we consider our rights, privileges and, above all, our duties.

(My Day, January 24, 1946)

We can't take it for granted that we are the only trustworthy people in the world and we must believe in other people's intelligence and good intentions if we expect them to believe in us.

(Letter to her daughter, Anna Roosevelt Boettiger, February 28, 1943)

Only if we come to understand the fears of other people which lead them to make certain demands, and their desire for economic or political power which necessitates certain other demands, will we be able to make each nation face its own situation in relation to similar situations in other nations.

(My Day, January 24, 1946)

In dealing with the new nations of the world we must relearn the meaning of that noble word *respect*. That is the only sound and enduring basis for any relationship among peoples, as it is among individuals.

(*Tomorrow Is Now*)

THE WORLD'S VIEW OF THE UNITED STATES

Our Declaration of Independence has become a docu-
ment read and studied by the whole world and it is not
too much to say that it is the gospel of hope. . . .

(My Day, July 4, 1946)

We should face the fact that to many of the nations of
the world the power of the United States is something
to be feared.

(My Day, January 26, 1951)

In a very real sense, the United States is the world's
show window of the democratic processes in action.

(*Tomorrow Is Now*)

An American traveling abroad is an ambassador not
only for the United States, but also for the concept of
democracy and, if he is a white American, for the entire
white race. Wherever he goes, America and democracy
will be thought of with a little more or a little less respect after he has departed.

(*Book of Common Sense Etiquette*)

No one likes the rich uncle who flaunts his wealth in the
face of your poverty; who will help you, perhaps—but on
his own terms; who will send you to college if you like—
but only to the college of his choice. This, of course, is

not a fair description of our attitude: but, nevertheless, fair or not, it is the way many see us.

(*India and the Awakening East*)

Every one of us in Europe or in the international meetings here knows how carefully everything that happens in the United States is watched today. No issue is any longer a completely domestic issue. It reaches the farthest corner of the world and is weighed neither for nor against us as a nation only but for or against democracy and our form of government and way of life.

(My Day, December 9, 1948)

AMERICA'S ROLE IN THE WORLD

There is hardly anything we undertake for the welfare of human beings—in medicine, in agriculture, in better teaching methods, in technicians who understand administration and organization, in the improvement and appreciation of crafts and the arts—which is not really part of what we can do for our defense and for the lifting of the morale of the people of the world.

(My Day, October 12, 1960)

If we were weak materially or spiritually, we might be afraid to try to understand the fears and prejudices and needs of other nations, and then we would hide our fear under threats of force. However, we are strong and we can afford to be conciliatory, though we can never afford to compromise our principles and our ideals.

(My Day, July 17, 1946)

We have an obligation, first of all, to solve our own problems at home, because our failure must of necessity, take away hope from the other nations of the world who have so much more to contend with than we have.

(Letter to Harry S. Truman, November 20, 1945)

The effort made in this country for freedom and control of government by the people was an inspiration to peoples everywhere. Now that we have grown strong and can help others attain the freedom they desire, we must see that it remains so.

(My Day, July 4, 1957)

We must not forget to look with care at the basic difficulties in the world. Starvation, lack of shelter, lack of clothing, lack of opportunity to enjoy life—these are the things that bring the world to the edge of doom today.

(My Day, January 1, 1951)

We cannot govern other people nor control what may happen in various parts of the world. We can only meet what happens as best we can, as it occurs, keeping always in mind our basic objective of peace.

(My Day, August 20, 1951)

In troubled areas of the world unpleasant things are going to happen under various forms of government. Only the big nations can give the example of self-control and calmness and a willingness to act along peaceful lines. This attitude is not a show of weakness.

(My Day, December 1, 1949)

We must move with calm moderation and speak softly even as we build our strength.

(My Day, January 29, 1951)

We, of course, hold paramount our own interests, but we know that our interests are best served when we consider those of the world as a whole.

No country in our position is loved, but we should make ourselves respected and understood.

(My Day, May 16, 1958)

Respect is not gained through weakness.

(My Day, April 16, 1957)

The more I traveled throughout the world the more I realized how important it is for Americans to see with understanding eyes the other peoples of the world whom modern means of communication and transportation are constantly making closer neighbors. Yet the more I traveled the happier I was that I happened to have been born in the United States, where there exist the concept of freedom and opportunities of advancement for individuals of every status. I felt, too, the great responsibility that has come to us as a people. The world is looking to us for leadership in almost every phase of development of the life of peoples everywhere.

(*The Autobiography of Eleanor Roosevelt*)

Today, we are one of the oldest governments in existence; ours has been the position for leadership, for setting the pattern of behavior. And yet we are supinely putting ourselves in the position of leaving the leadership to the Russians, of following their ideas rather than our own. For instance, when the Russians set up a restriction on what visitors to the country may be allowed to see, we promptly do the same thing here, in retaliation. Whenever we behave in this manner we are copying the methods of dictatorship and making a hollow boast of our claim that this country loves freedom for all. We owe it to ourselves and to the world, to our own dignity and self-respect, to set our own standards of behavior, regardless of what other nations do.

(*The Autobiography of Eleanor Roosevelt*)

We should live up to our own beliefs, giving others in many lands the opportunity of choosing between the products we produce, both cultural and material, and that which is produced by others who hold different views. Above everything, we must insist that we learn to live together in the future and that the primary aim of a nation is no longer to learn to die for one's country. It is more difficult, but far more necessary, to learn to live for one's country.

(My Day, February 27, 1961)

It is obvious that if we keep turning back we'll never move forward, and the new world will leave us behind. The arch conservatives want a shrinking and not an expanding America, with its sphere of influence ending at its own doorstep. And yet it must be plain that our world leadership may go by default if we practice nonintervention, or if we concentrate merely on fighting the Communists within, or if we continue our static thinking.

(Tomorrow Is Now)

By and large, the sum total of this nation's foreign policy is the combined voice of its people. Our impact on the rest of the world is the sum total of what each of us does as a private citizen. We tell citizens in foreign lands what kind of people we are by what we do here.

(Tomorrow Is Now)

The world is waiting for us to provide an example of dynamic drive, a bold reaffirmation of the values on which our nation was founded. Staying aloof is not a solution; it is a cowardly evasion.

(Tomorrow Is Now)

FOREIGN AID

Merely lending money to the world is not enough. Merely giving them a sense of military security and strength is not enough. What is really needed is inspiration and example.

(My Day, March 6, 1950)

The value of military aid is doubtful when given to nations that have not yet attained a sufficiently good standard of living to make life worthwhile.

(My Day, May 22, 1957)

I have always felt that women and children should be helped regardless of political ideas, and food and medical supplies should not be used as a political weapon.

(Letter to Harry Boardman, January 11, 1947)

Gratitude and love are not to be had for the asking; they are not to be bought. We should not want to think that they are for sale. What we should seek, rather than gratitude or love, is the respect of the world. This we can earn by enlightened justice. But it is rather naïve of us to think that when we are helping people our action is entirely unselfish. It is not.

(*The Autobiography of Eleanor Roosevelt*)

Twelve

THE WAY FORWARD

ELEANOR ROOSEVELT APPROACHED THE FUTURE WITH the same zest and energy she brought to every other area of her life. The fact that the future often seemed opaque or threatening did not frighten her. Instead the challenge of the unknown fed her curiosity, fired her imagination, and gave purpose to her life.

As early as 1934, she had recognized that America was "going through a period of transition. The doctrine [of] rugged individualism, widely held in this country, has been found inadequate for the needs of the future. We are groping for the next step, for a formula which meets our emerging needs."[1] Her formula for the future involved looking to the past for strength and inspiration. America's founders, she believed, had been driven "by a desire to create a new kind of civilization . . . where men could develop freely and fully their best abilities and capacities."

They had made their own history, and she believed her contemporaries could do the same. "The course of history is directed by the choices we make and our choices grow out of the ideas, the beliefs, the values, the dreams of the people," she wrote. "It is not so much the powerful leaders that determine our destiny as the more powerful influence of the combined voice of the people themselves."[2]

To those naysayers and fearmongers who wanted "to meet the future as though it were the past,"[3] she said pointedly, "People who 'view with alarm' never build anything."[4] Instead she challenged her fellow Americans to abandon their fears, draw courage from their past, and accept the responsibility of building "a free and peaceful world."[5]

THE AMERICAN DREAM

The American Dream can no more remain static than can the American nation. What I am trying to point out is that we cannot any longer take an old approach to world problems. They aren't the same problems. It isn't the same world. We must not adopt the methods of our ancestors; instead, we must emulate that pioneer quality in our ancestors that made them attempt new methods for a New World.

(The Autobiography of Eleanor Roosevelt)

No single individual, of course, and no single group has an exclusive claim to the American Dream. But we have all, I think, a single vision of what it is, not merely as a hope . . . but as a way of life, which we can come ever closer to attaining in its ideal form if we keep shining and unsullied our purpose and our belief in its essential value.

(The Autobiography of Eleanor Roosevelt)

The American Dream is never entirely realized. If many of our young people have lost the excitement of the early settlers who had a country to explore and develop, it is because no one remembers to tell them that world has never been so challenging, so exciting; the fields of adventure and new fields to conquer have never been so limitless.

(The Autobiography of Eleanor Roosevelt)

THE PAST

Today we are going through a period of transition. The doctrine rugged individualism, widely held in this country, has been found inadequate for the needs of the future. We are groping for the next step, for a formula which meets our emerging needs in this country. That is the responsibility and the chief interest of education today.

("A Message to Parents and Teachers,"
Progressive Education, January–February 1934)

For some people, of course, there is a nostalgic charm about certain ideas of the past. That idea of rugged individualism, completely divorced from the public interest, for example. It has a heroic sound, a kind of stalwart simplicity. The only trouble is that for many years it has been inapplicable to American life.

(*Tomorrow Is Now*)

We cannot turn back to a past historical era and attempt to live there. . . . What we need from our past history is to learn its lessons, profit by our mistakes, analyze our successes, find out all that it has to teach us.

(*Tomorrow Is Now*)

It is not enough to restore people to an old and outworn pattern. People must be given the chance to see the possibilities of a new world and to work for it.

(*My Day*, December 16, 1941)

We are prone always to think of the conditions which are with us today as being permanent conditions. To have a vision or a dream one must be able to guess at what changing conditions may bring and prepare for them.

(My Day, February 19, 1937)

THE FUTURE

Potential courage, knowledge, imagination, and vision are in our people, for they have had it in the past and they will certainly have it again if only their leaders will trust them, tell them the truth and draw from them the qualities that are needed to meet the crisis and the problems of the day.

(My Day, November 28, 1957)

It does not matter much whether we leave individually money or worldly goods of any kind, but it matters a great deal where we throw our weight in the development of society.

(*It's Up to the Women*)

Today we have achieved so much more, in many ways, than our ancestors imagined that sometimes we forget that they dreamed not just for us but for mankind.

(*The Autobiography of Eleanor Roosevelt*)

Every age, someone has said, is an undiscovered country. We are constantly advancing, like explorers, into the unknown, which makes life an adventure all the way. How interminable and dull that journey would be if it were on a straight road over a flat plain, if we could see ahead the whole distance, without surprises, without the salt of the unexpected, without challenge. I wish with all my heart that every child could be so imbued with a sense of the adventure of life that each change, each readjustment, each surprise—good or bad—that came along would be welcomed as a part of the whole enthralling experience.

(*You Learn by Living*)

Nothing of what has happened to me, or to anyone, has value unless it is a preparation for what lies ahead. We face the future fortified only with the lessons we have learned from the past.

(*Tomorrow Is Now*)

It is a wonderful thing to talk to people who have courage about the future.

(My Day, October 27, 1944)

I doubt if any human beings, just by themselves, are very important; but when they start a new trend of thought and action, they are apt to symbolize for their contemporaries and for the future a new idea, and therefore they become important.

(If You Ask Me, *Ladies' Home Journal,* August 1942)

Surely no one can seriously take the position that political ideas suited to a struggling young nation, shut
off from the world by wide oceans and connected to it
only by leisurely sailing ships, apply with any validity
to a powerful nation conditioned to men in orbit and a
world that can be encircled in minutes.

This seems as pointless as to suggest that we revert
to the horse and buggy because Washington traveled
that way.

(*Tomorrow Is Now*)

Sometimes, seeing the stubborn resistance of large
groups of Americans to accepting the existence of totally
new conditions, their determination to meet the future
as though it were the past, I am deeply puzzled. How did
it happen that a people with constantly developing ideas
on methods of production and distribution appears unable to develop new ideas, new points of view, new solutions to the problem of adjustment to change?

(*Tomorrow Is Now*)

The world must move and . . . when it moves, some of
the things we have cherished in the past are bound to be
destroyed. Perhaps we can build up some things that are
even better if we have the courage and belief in the future.

(*My Day, May 18, 1946*)

Person after person has said to me in these last few
days that this new world we face terrifies them. I can

understand how that feeling would arise unless one believes that men are capable of greatness beyond their
past achievements.

(My Day, August 10, 1945)

I can think of a thousand things in the past for which I
am deeply thankful, but it is for the future really that I
am most grateful—for the chance to try again to build
a decent world; for the young people who are so much
better educated in world affairs than we were twenty-
odd years ago, and who have high hopes and visions,
but who stand foursquare and face the realities of life.

(My Day, November 26, 1942)

What, indeed, may happen to us, depends upon our
capacity to think afresh, to understand the historical
trends, and to plan in accordance with what is actually
happening, not what we would prefer to have happening.

(*Tomorrow Is Now*)

There is no conceivable excuse for a great nation like
ours to be carried along helplessly on "the wave of the
future." We can decide the direction of our own course.

(*Tomorrow Is Now*)

It is not wishful thinking that makes me a hopeful
woman. Over and over, I have seen, under the most im-
probable circumstances, that man can remake himself,
that he can even remake his world if he cares enough

to try. And I have seen him, by the dozen, by the thousands, making that effort. Given leverage enough, a wise man said, "I could lift the world." Given incentive enough, man can remake his world. The incentive is his own well-being, the opportunity to grow to his full stature. Little by little, he is coming to know that and to grope for the point of leverage.

(*You Learn by Living*)

The future is literally in our hands to mold as we like. But we cannot wait until tomorrow. Tomorrow is now.

(*Tomorrow Is Now*)

Acknowledgments

THE EDITOR THANKS THE LITERARY ESTATE OF ELEANOR ROOSEVELT and the estate's executor, Nancy Roosevelt Ireland, for permission to publish excerpts from Eleanor Roosevelt's work and the Eleanor Roosevelt Papers Project for permission to use the digital editions of My Day and If You Ask Me. The editor also gratefully acknowledges the support of Liza Dawson and Anna Olswanger, who represent the literary estate, as well as Christopher Brick and Dr. Christy Regenhardt of the Eleanor Roosevelt Papers Project. She would also like to thank HarperCollins for permission to publish quotes from the following books:

You Learn by Living
The Autobiography of Eleanor Roosevelt

Notes

INTRODUCTION

1. Mary Ann Glendon, *A World Made New: Eleanor Roosevelt and the Universal Declaration of Human Rights* (New York: Random House, 2002), xxi.
2. Eleanor Roosevelt, Address to Americans for Democratic Action, April 1, 1950, reprinted in Allida M. Black, ed., *Courage in a Dangerous World: The Political Writings of Eleanor Roosevelt* (New York: Columbia University Press, 1999), 259–63.
3. Eleanor Roosevelt, *Tomorrow Is Now* (New York: Penguin Books, 2012), 5.
4. Eleanor Roosevelt, *You Learn by Living: Eleven Keys for a More Fulfilling Life* (Louisville, KY: Westminster Press, 1960), 41.
5. Roosevelt, *Tomorrow Is Now*, 22.
6. Eleanor Roosevelt, "In Defense of Curiosity," *Saturday Evening Post*, August 24, 1935. Available from the Eleanor Roosevelt Papers Project (hereafter ERPP), Columbian College of Arts and Sciences, https://erpapers.columbian.gwu.edu /defense-curiosity (accessed January 28, 2019).

CHAPTER 1: POLITICS AND GOVERNMENT

1. Roosevelt, *Tomorrow Is Now*, 110.
2. Ibid., 20.
3. Roosevelt, *You Learn by Living*, 194–95.
4. Eleanor Roosevelt, If You Ask Me, *McCall's*, April 1961. All articles from the If You Ask Me column available from ERPP, https://erpapers.columbian.gwu.edu/if-you-ask-me-0 (accessed January 28, 2019).
5. Roosevelt, *You Learn by Living*, 189.

CHAPTER 2: RACE AND ETHNICITY

1. Eleanor Roosevelt, "Keepers of Democracy," *Virginia Quarterly Review* 15 (January 1939): 1–5. Available from ERPP, https://erpapers.columbian.gwu.edu/keepers-democracy (accessed January 28, 2019).
2. Eleanor Roosevelt, "Tolerance Is an Ugly Word," *Coronet*, July 1945.
3. Eleanor Roosevelt, My Day, December 16, 1941. All articles from the My Day column available from ERPP, https://erpapers.columbian.gwu.edu/browse-my-day-columns (accessed January 28, 2019).
4. Roosevelt, My Day, June 19, 1943.

CHAPTER 3: FREEDOM AND RIGHTS

1. Roosevelt, If You Ask Me, *McCall's*, November 1960.
2. Roosevelt, My Day, April 15, 1943.
3. "A Challenge to American Sportsmanship," *Collier's*, October 16, 1943. Available from ERPP, https://erpapers.columbian.gwu.edu/challenge-american-sportsmanship (accessed January 28, 2019).
4. Roosevelt, My Day, August 29, 1952.
5. Eleanor Roosevelt, "Freedom of Speech," October 14, 1941, quoted in Tamara K. Hareven, *Eleanor Roosevelt: An American Conscience* (Chicago: Quadrangle Books, 1968), 165.

CHAPTER 4: ECONOMICS

1. Roosevelt, *Tomorrow Is Now*, 43.
2. Eleanor Roosevelt, *If You Ask Me* (New York: D. Appleton-Century, 1946), 35.
3. Eleanor Roosevelt, *The Moral Basis of Democracy* (New York: Open Road Integrated Media, 2016), 70.
4. Roosevelt, *Tomorrow Is Now*, 35.

CHAPTER 5: WOMEN, MEN, AND GENDER

1. Lois Scharf, "Equal Rights Amendment," in Maurine Beasley, Holly C. Shulman, and Henry R. Beasley, eds., *The Eleanor Roosevelt Encyclopedia* (Westport, CT: Greenwood Press, 2001), 161–65.
2. Eleanor Roosevelt, "Women in Politics," *Good Housekeeping*, March and April 1940.
3. Ibid.

CHAPTER 6: YOUTH AND EDUCATION

1. Roosevelt, My Day, January 14, 1953.
2. Roosevelt, *Tomorrow Is Now*, 70–71.

CHAPTER 7: FAITH AND ETHICS

1. James Chowning Davies, "Religion: Growth In," in Beasley, Shulman, and Beasley, eds., *The Eleanor Roosevelt Encyclopedia*, 437–40.
2. Roosevelt, If You Ask Me, *Ladies' Home Journal*, October 1941.
3. Roosevelt, My Day, July 5, 1962.

CHAPTER 8: LIFE LESSONS

1. Roosevelt, *You Learn by Living*, 1.
2. Roosevelt, My Day, February 4, 1936, and August 16, 1941; Roosevelt, *You Learn by Living*, 111.
3. Roosevelt, My Day, October 2, 1946.

CHAPTER 9: EMOTIONS AND HABITS

1. Roosevelt, *You Learn by Living*, 25, 41, 167–68.
2. Roosevelt, *Tomorrow Is Now*, 77; *You Learn by Living*, 30–31; *The Autobiography of Eleanor Roosevelt* (New York: Harper Perennial, 2014), 412.

CHAPTER 10: WAR AND PEACE

1. Roosevelt, *Tomorrow Is Now*, 98.
2. Roosevelt, My Day, December 9, 1942.

CHAPTER 11: THE UNITED STATES AND THE WORLD

1. Eleanor Roosevelt, *India and the Awakening East* (New York: Harper & Brothers, 1953), 114.
2. Roosevelt, My Day, January 7, 1950, and September 28, 1945.
3. Roosevelt, *Tomorrow Is Now*, 22.
4. Ibid., 46.

CHAPTER 12: THE WAY FORWARD

1. "A Message to Parents and Teachers," *Progressive Education*, January–February 1934.
2. Roosevelt, *Tomorrow Is Now*, 10.
3. Ibid., 22.
4. Ibid., 5.
5. Ibid., 128.

Bibliography

Eleanor Roosevelt Papers = *ERP*
Eleanor Roosevelt Papers Project = ERPP

BOOKS

Allison, Jay, and Dan Gediman, eds. *This I Believe: The Personal Philosophies of Remarkable Men and Women.* New York: Henry Holt, 2006.

Beasley, Maurine, ed. *The White House Press Conferences of Eleanor Roosevelt.* New York: Garland, 1983.

Beasley, Maurine, Holly C. Shulman, and Henry R. Beasley, eds. *The Eleanor Roosevelt Encyclopedia.* Westport, CT: Greenwood Press, 2001.

Binker, Mary Jo, ed. *If You Ask Me: Essential Advice from Eleanor Roosevelt.* New York: Atria, 2018.

Black, Allida, ed. *What I Hope to Leave Behind: The Essential Essays of Eleanor Roosevelt.* Brooklyn, NY: Carlson Publishing, 1995.

Black, Allida, et al., eds. *The Eleanor Roosevelt Papers.* Vol. 1, *The Human Rights Years, 1945–1948.* Farmington Hills, MI: Cengage Gale, 2007.

————, eds. *The Eleanor Roosevelt Papers*. Vol. 2, *The Human Rights Years, 1949–1952*. Charlottesville: University of Virginia Press, 2012.

Glendon, Mary Ann. *A World Made New: Eleanor Roosevelt and the Universal Declaration of Human Rights*. New York: Random House, 2002.

Hareven, Tamara K. *Eleanor Roosevelt: An American Conscience*. Chicago: Quadrangle Books, 1968.

Lash, Joseph P. *Eleanor: The Years Alone*. New York: W. W. Norton & Co., Inc., 1972.

————. *Love, Eleanor: Eleanor Roosevelt and Her Friends*. Garden City, NY: Doubleday, 1982.

————. *A World of Love: Eleanor Roosevelt and Her Friends, 1943–1962*. Garden City, NY: Doubleday, 1984.

Roosevelt, Eleanor. *The Autobiography of Eleanor Roosevelt*. New York: Harper Perennial, 2014.

————. *Eleanor Roosevelt's Book of Common Sense Etiquette*. New York: Macmillan, 1962.

————. *If You Ask Me*. New York: D. Appleton-Century, 1946.

————. *India and the Awakening East*. New York: Harper & Brothers, 1953.

————. *It's Up to the Women*. New York: Nation Books, 2017.

————. *The Moral Basis of Democracy*. New York: Open Road Integrated Media, 2016.

————. *This Troubled World*. New York: H. C. Kinsey, 1938.

————. *Tomorrow Is Now*. New York: Penguin Books, 2012.

————. *You Learn by Living: Eleven Keys for a More Fulfilling Life*. Louisville, KY: Westminster Press, 1960.

Smith, Stephen Drury. *The First Lady of Radio: Eleanor Roosevelt's Historic Broadcasts*. New York: The New Press, 2014.

Woloch, Nancy, ed. *Eleanor Roosevelt: In Her Words: On Women, Politics, Leadership, and Lessons from Life*. New York: Black Dog & Leventhal, 2017.

CORRESPONDENCE AND OTHER DOCUMENTS

From the Anna Eleanor Roosevelt Papers, Franklin D. Roosevelt Library

Diary Entry, London, January 9, 1946. (Also found in *ERP*, vol. 1, 195–96.)

Letter to Harry Boardman, January 11, 1947. (Also found in *ERP*, vol. 1, 468.)

Letter to Elsa Marcussen, February 2, 1948. (Also found in *ERP*, vol. 1, 733.)

Letter to Margaret Carberry, February 22, 1949. (Also found in *ERP*, vol. 2, 41–42.)

Letter to Francis Cardinal Spellman, July 23, 1949. (Also found in *ERP*, vol. 1, 169–71.)

Letter to L. M. Deemer, August 29, 1949. (Also found in *ERP*, vol. 2, 193–94.)

Letter to Sara Gould, January 19, 1951. (Also found in *ERP*, vol. 2, 526–27.)

Letter to Francis Horn, February 19, 1960.

Letter to Anna Roosevelt Boettiger, February 28, 1943, quoted in *Mother and Daughter: The Letters of Eleanor and Anna Roosevelt*, edited by Bernard Asbell (New York: Fromm International, 1988), 156–57.

Undated draft statement circa 1951 for *Red Tape: The Civil Service Magazine*. (Also found in *ERP*, vol. 2, 777.)

From the Harry S. Truman Presidential Library Online

Letter to Harry S. Truman, November 20, 1945, https://truman library.org/eleanor/1945.html.

Memo for the President, March 1, 1946, https://trumanlibrary .org/eleanor/1946.html.

Letter to Harry S. Truman, June 30, 1946, https://trumanlibrary .org/eleanor/1946.html.

Memo for the President, December 28, 1948, https://truman library.org/eleanor/1948.html.

Letter to Harry S. Truman, January 29, 1952, https://truman library.org/eleanor/1952.html.

ARTICLES

Online

"A Challenge to American Sportsmanship," *Collier's*, October 16, 1943. Available at ERPP, https://erpapers.columbian.gwu.edu /challenge-american-sportsmanship.

"Good Citizenship: The Purpose of Education," *Pictorial Review*, April 1930. Available at ERPP, https://erpapers.columbian .gwu.edu/good-citizenship-purpose-education.

"How to Take Criticism," *Ladies Home Journal*, November 1944. Available at ERPP, https://erpapers.columbian.gwu.edu/how -take-criticism.

"I Want You to Write to Me," *Women's Home Companion*, August 1933. Available at ERPP, https://erpapers.columbian.gwu.edu /i-want-you-write-me.

"If I Were A Republican Today," *Cosmopolitan*, June 1950. Available at ERPP, https://erpapers.columbian.gwu.edu/if-i -were-republican-today.

If You Ask Me. Available at ERPP, https://erpapers.columbian .gwu.edu/if-you-ask-me-0.

"In Defense of Curiosity," *Saturday Evening Post*, August 24, 1935. Available at ERPP, https://erpapers.columbian.gwu.edu /defense-curiosity.

"Keepers of Democracy," *Virginia Quarterly Review* 15 (January 1939). Available at ERPP, https://erpapers.columbian.gwu .edu/keepers-democracy.

"Liberals in This Year of Decision," *Christian Register*, 127 June 1948. Available at ERPP, https://erpapers.columbian.gwu .edu/liberals-year-decision.

My Day. Available at ERPP, https://erpapers.columbian.gwu .edu/browse-my-day-columns.

"Race, Religion and Prejudice," *New Republic*, May 11, 1942. Available at ERPP, https://erpapers.columbian.gwu.edu/race -religion-and-prejudice.

"What I Hope to Leave Behind," *Pictorial Review*, April 1933. Available at ERPP, https://erpapers.columbian.gwu.edu/what -i-hope-leave-behind.

"What Religion Means to Me," *Forum* 88 (December 1932). Available at ERPP, https://erpapers.columbian.gwu.edu/what -religion-means-me.

"Why I Do Not Choose to Run," *Look*, July 9, 1946. Available at ERPP, https://erpapers.columbian.gwu.edu/why-i-do-not-choose-run.

Print

"The Elementary Teacher as a Champion of Human Rights," *The Instructor*, September 1951.

"Fear Is the Enemy," *The Nation*, February 10, 1940.

The Fears of Free Americans, student council pamphlet, March 26–28, 1954.

"General Assembly Adopts Declaration of Human Rights: Statement," *Department of State Bulletin* 19, no. 494 (December 19, 1948), 751–52.

"From the Melting Pot—An American Race," *Liberty*, July 14, 1945, reprinted in *ERP*, vol. 1, 66–70.

"How to Choose a Candidate," *Liberty*, November 5, 1932.

"In Service of Truth," *The Nation*, July 9, 1955.

"Insuring Democracy," *Collier's*, June 15, 1940, reprinted in *Courage in a Dangerous World: The Political Writings of Eleanor Roosevelt*, edited by Allida M. Black (New York: Columbia University Press, 1999), 71–75.

"Intolerance," *Cosmopolitan*, February 1940, reprinted in *Courage in a Dangerous World: The Political Writings of Eleanor Roosevelt*, edited by Allida M. Black (New York: Columbia University Press, 1999), 120–25.

"Let Us Have Faith in Democracy," *Land Policy Review* 5 (January 1942), 20.

"A Message to Parents and Teachers," *Progressive Education*, January–February 1934.

"A Place for Women in Politics," *Women's Home Companion*, 1932, reprinted in *Eleanor Roosevelt: In Her Words: On Women, Politics, Leadership, and Lessons from Life*, edited by Nancy Woloch (New York: Black Dog & Leventhal, 2017), 51.

"Reply to Attacks on U.S. Attitude toward Human Rights Covenant," *Department of State Bulletin* 26, no. 655 (January 14, 1952): 59.

"Segregation," *Educational Forum* 24, no. 1 (November 1959), 5–6, reprinted in *What I Hope to Leave Behind: The Essential Essays of Eleanor Roosevelt*, edited by Allida M. Black (Brooklyn, NY: Carlson Publishing, 1995), 179–80.

"Should Wives Work?," *Good Housekeeping*, December 1937.

"Social Gains and Defense," *Common Sense*, March 1941, reprinted in *Courage in a Dangerous World: The Political Writings of Eleanor Roosevelt*, edited by Allida Black (New York: Columbia University Press, 1999), 132–35.

"The State's Responsibility for Fair Working Conditions," *Scribner's Magazine*, March 1933.

"Ten Rules for Success in Marriage," *Pictorial Review*, December 1931.

"Tolerance Is an Ugly Word," *Coronet*, July 1945.

"What I Think of the United Nations," *United Nations World Magazine*, August 1949, reprinted in *Eleanor Roosevelt: In Her Words: On Women, Politics, Leadership, and Lessons from Life*, edited by Nancy Woloch (New York: Black Dog & Leventhal, 2017), 228.

"Why I Am Opposed to 'Right to Work' Laws," *AFL-CIO American Federationist*, February 1959.

"Women in Politics," *Good Housekeeping*, January, March, and April 1940.

ARTICLES ABOUT ELEANOR ROOSEVELT

"Eleanor Roosevelt—Things She Has Said," *Australian Women's Weekly*, September 11, 1943, https://ia802506.us.archive.org/14

/items/The_Australian_Womens_Weekly_11_09_1943/The
_Australian_Womens_Weekly_11_09_1943.pdf.
"Suspicion as Peace Bar Feared by Mrs. Roosevelt," *Newark Evening News*, October 2, 1945. (*ERP*, vol. 1, 112.)

BROADCASTS

Americans of Tomorrow, CBS, November 11, 1934. Transcript in
The First Lady of Radio: Eleanor Roosevelt's Historic Broadcasts,
edited by Stephen Drury Smith (New York: The New Press,
2014), 62–67 (hereafter cited as *FLR*).

The Eleanor Roosevelt Program, NBC, August 23, 1951, Anna Eleanor Roosevelt Papers, Franklin D. Roosevelt Library. (*ERP*,
vol. 2, 697.)

Interview by Mike Wallace, November 23, 1957, Anna Eleanor
Roosevelt Papers, Franklin D. Roosevelt Library.

Interview by Milton Cross, *The Magic Key of RCA*, NBC Blue
Network, January 9, 1938. Transcript in *FLR*, 114–20.

Mrs. Eleanor Roosevelt's Own Program, NBC Red Network,
June 20, 1940. Transcript in *FLR*, 136–40.

Over Our Coffee Cups, NBC Blue Network, December 7, 1941, and
February 15, 1942. Transcripts in *FLR*, 189–97, 210–15.

The Pond's Program, February 3, 1933, and May 5, 1937, NBC Red
Network. Transcripts in *FLR*, 29–39, 94–100.

The Simmons Program, NBC Blue Network, September 18, 1934.
Transcript in *FLR*, 51–55.

SPEECHES

Address at Stuttgart, Germany, October 23, 1948, Anna Eleanor
Roosevelt Papers, Franklin D. Roosevelt Library. (*ERP*, vol. 1,
924–26.)

Address to Americans for Democratic Action, Washington,
DC, April 1, 1950, reprinted in *Courage in a Dangerous
World: The Political Writings of Eleanor Roosevelt*, edited by

Allida M. Black (New York: Columbia University Press, 1999), 259–63.

Address to the Democratic National Convention, Chicago, IL, July 22, 1952, Anna Eleanor Roosevelt Papers, Franklin D. Roosevelt Library. (*ERP*, vol. 2, 907–11.)

Address to Les Jeune Amis de la Liberté, Paris, France, December 18, 1951, Anna Eleanor Roosevelt Papers, Franklin D. Roosevelt Library. (*ERP*, vol. 2, 736)

Address to Roosevelt College, Chicago, IL, November 6, 1945, Roosevelt University Archives. (*ERP*, vol. 1, 136–38.)

Campaign address for Adlai Stevenson, Charleston, WV, October 1956, Anna Eleanor Roosevelt Papers, Franklin D. Roosevelt Library.

"Civil Liberties—The Individual and the Community," address to the Chicago Civil Liberties Committee, *Reference Shelf* 14 (March 14, 1940), 173–82, reprinted in *What I Hope to Leave Behind: The Essential Essays of Eleanor Roosevelt*, edited by Allida M. Black (Brooklyn, NY: Carlson Publishing, 1995), 149–54.

"Negro Education," speech to the National Conference on the Education of Negroes, May 11, 1934, in *FLR*, 36–41.

Remarks at the meeting of the UN General Assembly Third Committee, October 28, 1947.

Speech at the Phi Beta Kappa Association Founders' Day Dinner, Chicago, IL, February 29, 1949. Available at ERPP, https://er papers.columbian.gwu.edu/speech-phi-beta-kappa-association -founders-day-dinner-1949.

Statement for Paris, April 14, 1949, Anna Eleanor Roosevelt Papers, Franklin D. Roosevelt Library. (*ERP*, vol. 2, 70.)

"The Struggle for Human Rights," speech at the Sorbonne, Paris, France, September 28, 1948. Available at ERPP, https://erpapers .columbian.gwu.edu/struggle-human-rights-1948.

"Where Do Human Rights Begin?," remarks at the United Nations, March 27, 1958, quoted in Joseph P. Lash, *Eleanor: The Years Alone* (New York: W. W. Norton, 1972), 81.

"Why the United Nations Is Unpopular—And What We Can Do about It," speech to Citizens Conference on International Economic Union, November 19, 1952, Anna Eleanor Roosevelt Papers, Franklin D. Roosevelt Library. (*ERP*, vol. 2, 972–76.)

About the Contributors

MARY JO BINKER IS A CONSULTING EDITOR FOR THE ELEANOR Roosevelt Papers Project at George Washington University and an adjunct professor of history at George Mason University. In addition to her work at the Project, she edited *If You Ask Me: Essential Advice from Eleanor Roosevelt*, a selection of questions and answers drawn from the magazine column of the same name. Her articles have appeared in *Time* (online), *Ms.* (online), and *White House History Quarterly*. She holds a bachelor's degree in communication from Oregon State University and a master's degree in history from George Mason University.

NANCY PELOSI IS THE FIFTY-SECOND SPEAKER OF THE HOUSE OF Representatives, having made history in 2007 when she was elected the first woman to that post. Now in her third term as Speaker, Pelosi made history again in January 2019 when she regained her position second in line to the presidency, the first person to do so in more than sixty years.

Pelosi brings to her leadership position a distinguished record of legislative accomplishment. She led the Congress in passing historic health insurance reform, key investments in college aid, clean energy and innovation, and initiatives to help small businesses and veterans. Married to Paul Pelosi, she is a mother of five and grandmother of nine.

About the Author

IN HER TWELVE YEARS AS FIRST LADY (1933–1945), ELEANOR Roosevelt transformed that role, traveling extensively at home and abroad, holding regular press conferences, hosting radio programs, writing newspaper and magazine columns, and championing the rights of women, minorities, youth, and labor. After leaving the White House in 1945, she became a diplomat, serving as a delegate to the United Nations General Assembly (1946–1952) and as the first chair of the UN Commission on Human Rights. In the latter role, she was instrumental in the creation and passage of the Universal Declaration of Human Rights (1948). Domestically, she also continued to be active in the civil rights, labor, and women's movements as well as in the Democratic Party. In 1961 President John F. Kennedy reappointed her to the UN and also named her chair of the President's Commission on the Status of Women, a position she held until her death in 1962. In addition to her newspaper and magazine columns and articles, she wrote twenty-seven books, four of which are autobiographical: *This Is My Story, This I Remember, On My Own,* and *The Autobiography of Eleanor Roosevelt.*

BOOKS BY ELEANOR ROOSEVELT

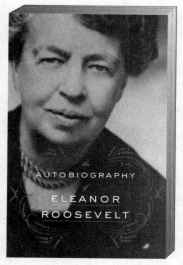

YOU LEARN BY LIVING
Eleven Keys for a More Fulfilling Life

Available in Paperback and eBook

One of the world's best loved and most admired public figures, Eleanor Roosevelt remains a role model for a life well lived. *You Learn by Living* distills her philosophy and outlines the strategies she used to create a rich and meaningful life for herself and others. Based on her own hard-won experience, *You Learn by Living* offers practical ways to cultivate compassion, overcome fears, embrace challenges, and become a more informed, engaged citizen. Filled with warmth, common sense, and realistic optimism, *You Learn by Living* is a window into Eleanor Roosevelt's life and a trove of timeless wisdom.

THE AUTOBIOGRAPHY OF ELEANOR ROOSEVELT

Available in Paperback and eBook

The Autobiography of Eleanor Roosevelt is a candid, insightful look at an era and a life written by one of the most remarkable Americans of the twentieth century. Against the backdrop of some of the most dramatic events in U.S. history including two world wars, the Great Depression, and the Cold War, she traces her own development from a shy, awkward orphan to a political and social activist, journalist, transformative First Lady, and ultimately, an internationally recognized champion of human rights. Apart from her insights into the events she lived through and the leaders she interacted with, what makes Eleanor Roosevelt's autobiography compelling is her courage, her zest for life, and her profound conviction that everyone can make a contribution and everyone should try.